Isaac Shinn

The Ready Adviser and Family Guide

A New Compilation of Valuable Recipes and Guide to Health with Directions What to Do in Cases of Emergency

Isaac Shinn

The Ready Adviser and Family Guide
A New Compilation of Valuable Recipes and Guide to Health with Directions What to Do in Cases of Emergency

ISBN/EAN: 9783337811778

Printed in Europe, USA, Canada, Australia, Japan

Cover: Foto ©berggeist007 / pixelio.de

More available books at **www.hansebooks.com**

THE
READY ADVISER

AND

FAMILY GUIDE.

A NEW COMPILATION OF

VALUABLE RECIPES AND GUIDE TO HEALTH;

WITH

DIRECTIONS WHAT TO DO IN CASES OF EMERGENCY;

COMPRISING OVER

ONE THOUSAND VALUABLE RULES AND RECIPES USEFUL TO EVERY BODY,

AND DIVIDED INTO

FOUR PARTS,

WITH A FULL INDEX FOR EACH PART.

By ISAAC SHINN, Esq.

CHICAGO:
CHURCH & GOODMAN, 51 LaSALLE STREET.
1866.

PREFACE.

In presenting this BOOK OF RECIPES to the public, I hope to receive that indulgence due to one who acts under the advice and at the solicitation of others, as well as under his own conviction of duty.

Having been a sufferer myself, I can readily sympathize with the afflicted; and as a number of these recipes, which are known to be almost invaluable, having never before been made public, I feel that I am but simply discharging my duty in publishing them.

The greater number of them have been gathered from various sources, through the space of thirty years; to which I have added a few of my own that are known to be good.

A number of recipes that have fallen into my hands, have been rejected as worthless, while others have been inserted which, doubtless, should have met the same fate.

I have endeavored, however, to make the book not only reliable, but to so arrange its contents, that almost any recipe can be readily found. In doing this, I have divided the book into four parts, and each part has again been divided into as many different headings as it contains subject matters; and under each heading I have given all the recipes that belong to that given subject. This rule has been observed throughout, excepting in the case of a few miscellaneous items.

I have usually given several recipes on the same subject, so that if any should not have the ingredients of one, but should have of others, they might know how to apply that which they have at hand.

PREFACE.

The Index refers to the page containing the first recipe on any given subject, and all the recipes on that subject are continued consecutively from that page on, until all are given.

I have added a Medical Flora, in which I have given some of the properties and uses of the most common of the medical trees, shrubs, plants, herbs, and flowers usually found in this country.

Thirty years having elapsed since gathering some of these recipes, the sources whence they were derived are partly forgotten, or I would make that acknowledgment I otherwise most certainly should.

Having devoted a good deal of time, with some expense, in gathering and writing these recipes, and having compiled them with that care and attention that such a work requires, I now submit it to a generous public, with the hope that they will make that allowance due to one whose aim has been more to promote the good of his fellow-man, than for his own reward.

THE AUTHOR.

QUINCY, ILL., April 10th, 1866.

INDEX.

MEDICAL DEPARTMENT.

	PAGE.		PAGE.
Artery or Vein, Cut,	21	Boneset,	139
Antidotes for Poisons,	23-27	Burdock,	139
Arsenic, Antidote for	24-25		
Alcohol, " "	25	Clothing,	17-19
Alkali, " "	27	Clothes taking fire, what do,	22-23
Ammonia, " "	27	Corrosive Sublimate, Antidote for	26
Articles of Diet, time in digesting,	56-57	Creosote, Antidote for	26
Asthma,	138	Carbonic gas, "	27
Ague and Fever,	113-115	Cholera,	37-45 and 236
Ague, Dumb, etc.,	115	Checking Perspiration,	52-54
Apoplexy and Paralysis,	122	Consumption,	76-80
Advice to Mothers,	138	Coughs and Colds,	80-83
Alder, its uses,	138	Croup,	83-86
		Cancer,	86-87
Bathing,	16	Cancer Wart,	87
Belladonna, Antidote for	26	Colic,	98-99
Bleeding from Vein or Artery,	21 and 96	Colic or Cramp in the Stomach,	99
Bleeding from Wounds,	94-95	Chills and Fever,	113-115
Bleeding from the Nose,	95-96	Catarrh in the Head,	117
Bleeding from the Lungs or Stomach,	96	Chapped Hands,	120
		Children, Feeding of	121
Breasts, Sore and Hard,	99-100	Corns,	122-123
Burns and Scalds,	104	Cramp,	124
Bronchitis,	115	Chillblains,	124
Brain, Inflammation of	116	Court Plaster,	125
Boils and Styes,	124	Cuts, Wounds, etc.,	126
Brandy, Properties of	130	Choked, on Fish-Bone	126
Balm of Gilead,	139	Cordial, Godfrey's	128
Blackberry,	139	Children, Physic for	129
Black Snake Root,	139	Composition Powders,	129

INDEX.

Coffee, Effects of Roasted.. 130
Catnip, ... 139
Comfrey, ... 139
Camomile, ... 139
Centaury, ... 139
Drinking, ... 13-14
Damp Houses, ... 15
Drowning persons, How to Save. 23
Deaths, Sudden 55-56
Digesting of Food, ... 56-57
Dyspeptics, what they sho'ld not eat, ... 57-58
Drowning, ... 59-60
Drowned, apparently, how treated, ... 59-60
Diphtheria, ... 63-68
Diarrhœa, ... 68-70
Dysentery, ... 71-73
Dyspepsia, ... 75-76
Drawing Salve, ... 107
Dropsy, ... 108
Dumb Ague, Weakness and General Debility, ... 115
Dislocation of the Jaw,.. 118
Dislocation (partial) of the Neck, ... 110
Diabetes in Children, ... 119-120
Diseases and Remedies, .. 121
Deafness, ... 122
Drunkenness, Cure for ... 131
Death, true signs of. 239
Dumb made to speak, ... 136
Dandelion, ... 139
Dittany, Mountain 140
Dogwood, ... 140
Eating, ... 13
Exercising, ... 16
Erysipelas, ... 74-75
Eyes, Sore and Weak ... 100-101
Earache, ... 103
Elecampane, ... 140
Elder, ... 140
Elm, Slippery 140

Food, time in digesting 56-57
Food, not good for Dyspeptics, ... 57-58
Food, for Fat and Lean People, ... 58-59
Felon, ... 87-89
Fever, Scarlet 112
Fever and Ague, ... 113-115
Frozen Limbs, ... 116
Fever Sore, ... 119
Fish Bone in Throat, ... 126
Flaxseed Syrup, how made, 128
Fennel Seed, ... 140
Guide to Health, ... 13-19
Gravel, ... 109-110
Goitre, ... 122
Gonorrhœa, ... 127
Gin, Properties of 131
Garlic, ... 140
Ginger, ... 141
Ginseng, ... 141
Ground Ivy, ... 141
Golden Seal, ... 141
House on fire, ... 20
Horses, Frightened 21
Horses, Balky, what do... 21
Harness, if they break, what do, ... 21
Horses, how to take from barn on fire, ... 21
Hydrophobia, ... 27-37
Hooping-cough 83
Headache, Nervous and Sick, ... 102
Hurts, Internal 106
Hiccough, ... 110
Hoarseness, ... 115-116
Hair, ... 137 and 153
Hops, ... 141
Horse-radish, ... 141
Hyssop, ... 141
Hoarhound, ... 141
Horse-mint, ... 141

INDEX.

	PAGE		PAGE
Inflammation, how to allay	22	Opium, Antidote for	26
Important for the Soldier,	45-52	Oxalic Acid, Antidote for	26
Itch, Cure for the	98	Old Sores and Ulcers,	105-106
Internal Hurts, or Ulcers,	106	Odor of Perspiration,	121
Influenza,	115	Onions, Efficacy of	129
Inflammation of the Brain,	116		
Inflammation of the Stomach	117	Presence of mind,	20-22
Inflammation of the Kidneys,	239	Persons drowning, What to do,	23
Indigestion,	127	Poisons and their Antidotes,	23-27
		Prussic Acid, Antidote for	26
Joints and Sinews	105	Perspiration, Checking	52-54
Jaw, Dislocation of	118	Persons apparently drowned, How to treat	59
Kidneys, Inflammation of the	239	Putrid Sore Throat,	90-91
		Phthisic,	101
		Piles,	104-105
Laudanum, Antidote for	26	Palsy, Numb	112
Lead, White and Sugar of, Antidote for	27	Palipus,	112
		Poison, Vegetable	120
Leprosy,	118	Paralysis and Apoplexy,	122
Lock-jaw,	120	Poisons in daily use,	135
Life Everlasting,	141	Parsley,	141
		Plantain,	142
Mushrooms, Antidote for	26	Prickly Ash,	142
Muriatic Acid, Antidote for	26	Pennyroyal,	142
Mortification,	125-126	Peppermint,	142
Mortification, Dry	126	Peruvian Bark,	142
Miasma, How to avoid	133		
Medical Flora,	138-143	Quinsy,	101
May-Apple,	141		
Mullen,	141	Rules for General Health,	45
		Rooms, how to purify	97-98
Nitric Acid, Antidote for	26	Rheumatism,	101-111
Nitrate of Potash, Antidote for	26	Ringworms,	123-124
		Rattle Weed, (See Black Snake Root,)	142
Nitrate of Silver, Antidote for	26		
Nux Vomica, Antidote for	26	Sleeping,	14
Neuralgia,	73-74	Sleeping Rooms,	14
Nipples and Breasts, Sore	99-100	Schooling,	19-20
Numb Palsy,	112	Sulphuric Acid, Antidote for	26
Neck, Partial dislocation of	119	Strychnine, Antidote for	26
Nose, to extract any substance from the Nostrils,	120	Snake-Bite, Antidote for	26
		Soldier, Information for the	45-50
Night Sweats,	127	Sudden Deaths	55-56
Number Six,	127	Small-pox,	60-63

	PAGE.		PAGE.
Snake-Bite, Cure for	89-90	Seneca Snake Root,	143
Sore Throat, (putrid)	90	Spikenard,	143
Salve, Healing	94 and 105		
Salve for Old Sores,	94 and 105	Taking Cold,	16-17
Sore Nipples and Breasts,	99-100	To Stop Bleeding,	21 and 95
Scalds and Burns,	104	Tobacco, Antidote for	26
Stiff Joints, etc.,	106	Tartar Emetic, Antidote for	27
Swellings,	107	Teas for the Sick,	96-97
Sprains,	107	Toothache,	103
Salve, Drawing,	107	Teeth,	124
Scarlet Fever,	112	Tonic.	128
Scarlatina and Measels,	113	Teas, Green and Black	128
Stomach, Inflammation of	117	Tansy,	143
Scrofula, etc.,	117-118		
Suffocation from Smoke,	125	Ulcer Powder,	94
Spider and Mosquito Bites,	125	Ulcer Salve,	94
String of Bees and Wasps,	125	Ulcer, Internal.	106
Sore Lips,	125	Upas Tree,	134
Sunstroke,	125		
Salve for Cuts, Wounds, etc.,	126	Visiting Sick Rooms,	15
Scald Head,	127	Verdigris, Antidote for	27
Scurf in the Head,	127	Vaccination,	54
Salivation,	127	Vomiting, How to Stop	239
Stammering,	127	Virginia Snake Root,	143
Sweats, Night	127		
Seidlitz Powders,	128	White Vitriol, Antidote for	27
Strychnine,	134	Woolen Clothing,	51-55
Signs of Death,	239	What Fat and Lean People	
Salt, Medical use of	138	can Eat, etc.,	58-59
Slippery Elm, (See Elm, Slippery,)	140	White Swelling,	91-94
		Worms in Children,	108
Sage,	142	Water, Stoppage of the	109-110
Sassafras,	142	Wen,	121
Sumach,	142	Warts,	123
Spearmint,	142	Wine, Properties of	130-131
Smart Weed,	142	Water Plantain,	143
Senna	143	Wild Cherry,	143
Sweet Fern,	143	Winter-green,	143
Sarsaparilla,	143	Yellow Root, (See Golden Seal,)	143

MISCELLANEOUS DEPARTMENT.

	PAGE.		PAGE.
Barometer, The Farmer's	144	Bed-bugs. Ants, Flies, etc.,	
Barometer, Leech	144	How to get rid of	149-150
Barometer, Cheap	145	Blacking for Boots,	151-152

INDEX.

	PAGE.		PAGE.
Cements,	147-148	To Sweeten Musty Casks,	154
Cloth from Moths,	150	To Prevent the Smoking of a Lamp,	154
Coffee Syrup,	161	To preserve a Bouquet,	154
Furs from Moths,	193	To soften Putty,	155
Flowers, Getting the perfume of	161	To make Lime Water,	155
Hair, How to Color Black,	153	To extinguish a Chimney on fire,	155
Hair, How to destroy superfluous	153	To break Glass in any shape,	156
		To renovate Cloths,	156
		To destroy foul smells,	156
Inks, How to make	152	To purify Cellars, etc.,	156
Leather, to preserve	151-152	To see to the bottom of cisterns, etc.,	154
		To make Liquid Glue,	155
Metals, to take rust from	148-149	To silver Ivory,	155
Paints,	146-147	To purify breath tainted by onions,	155
Plants, gathering the perfume of	161	To make Drawings resemble paintings in oil,	156
Rust from Metals,	148-149	To remove all kinds of stains from all kinds of cloth,	156-159
		To clean kid gloves,	159
Solder Liquid,	148	To dye Nankeen Color,	159
Stains, how to remove from all kinds of Clothing,	160	To dye Black,	159
		To dye Purple,	159
Swimming, Art of	160	To make Mucilage,	160
		To separate papers, etc.,	160
Tomato Worms,	151	To remove foul air from wells,	160
To Clean Flat-irons, 149 and	194		
To make wood fire or waterproof,	153	To brown Metals,	160
		To drive away Rats,	161
To make Paper or Cloth fire-proof,	153	Time Table,	162
To Renovate Manuscripts,	153	White-washes,	145-146

DOMESTIC DEPARTMENT.

	PAGE.		PAGE.
Bread, Breakfast, and Tea Cakes,	164	Blackberry Cordial,	210
		Brandy, Artificial	210
Butter, hard without ice,	190	Beers,	210-211
Butter, how to freshen salt	190		
Butter, how to preserve	190	Cookery, Domestic	164-187
Butter, how to pack	190	Catsup, Tomato	185-187
Butter, how to make yellow,	191	Coffees, how to make, etc.	187-189
Butter, good way to make	191	Churning,	190

INDEX.

	PAGE.		PAGE.
Cream, a substitute for....	191	Mildew on Trees, how to prevent	193
Candles from Lard,	193		
Coffee, Substitute for...	187-189	Moth from Furs, etc.,	193
Coffee as the French prepare it,	188	Meat, how to Cure, etc.,.	205-207
Cider, how to make and preserve,	207-208	Onions, how to sprout	185
		Oysters, Pickled	184
Cider, without Apples,....	208		
Cordial, Blackberry	210	Puddings and Pies,	178-181
		Potatoes, Frozen	192
Domestic Cookery,	164-187	Potatoes, how to cook	194
		Pickles and Preserves,..	197-203
Eggs, how to tell fresh....	194		
Eggs, how to preserve	194	Rum, Artificial	210
Fish, Salt, how to freshen..	184	Side Dishes,	181
Fish, how to Boil	184	Soda Water and Lemonade,	192
Flour, how to select good..	195	Sealing-wax for Fruit Cans,	203
		Sausages, how to make and keep	207
Gin, Artificial	210		
		Soap, how to make	112-113
Honey, Artificial	191		
Herbs, when to gather, etc.,	193	Tomato Catsup,	185-187
Hens, to make them lay,...	196	Tea, how to draw, etc.,	189
Hens, to make them lay all winter,	196	To restore wilted flowers,..	194
		To mend cracks in stoves,..	194
Hen Cholera,	196	To take rust off Flat-irons,.	194
		To keep Ants from closets, etc.,	195
Ice, how to keep from melting,	101	To destroy Flies, etc.,	195
		Tomatoes as Food,	186-187
Jars, how to clean the inside of	196	Vinegar, how to make...	211-212
		Warning to Housekeepers,.	186
Knives, how to clean	192	Water, how to make pure, cold and soft,	203-205
Kitchen Smells, how to avoid	194		
		Wines,	208-210
Lard, how to make	192	Washing, etc., etc.,	212-217
		Welsh Rabit,	184
Milk, how to preserve...	191-192	Yeast,	187

FARMERS' DEPARTMENT.

	PAGE.		PAGE.
Botts in Horses,	219	Cough Ball for Horses,	222
Colic in Horses,	219	Cathartic Powder,	223
Cough in Horses,	221	Cattle, choked, how to relieve	224

INDEX.

	PAGE.		PAGE.
Cattle, Physic for	225	Measures and Weights,	233-234
Cattle, weight by measure,	240		
Calves, Diarrhœa in	225	Poll-Evil,	222
Corn, Germinating	225	Physic Ball,	222
Chinch Bug,	226	Potato Rot Preventive,	226
Cholera,	236-237	Purgative Ball,	223
Diseases of Horses,	219-223	Rats, how to get rid of	229
Disinfectants, and how to use them,	234-235	Raising double Crops,	126
		Stifle,	220
Fistula in Horses,	222	Scratches,	220
Founder in Horses,	221		
Fever Ball for Horses,	223	Tonic for Horses and Cattle,	222
		To protect Grain, Fruit and Plants from Insects,	227
Glanders in Horses,	221		
Grains, Grass, etc.,	225-226	To make Peaches grow without Stones,	227
Grafting,	227		
Grains, etc., Weight of	233	Trees girdled by Mice or Rabbits,	228
Horses, Diseases of	219-223	To protect Fruit Trees from Curculio,	228
Horses, to keep flies from	224		
Horse, how to start a balky	224	To protect Peach Trees from Worms.	228
Hog Cholera	229		
Hog Itch,	229	To bring Dead Trees to Life,	228
		Timber, when to cut	229
Insect Trap,	227	To keep Tires on wheels,	229
Liniment for Blistering Horses,	223	To measure Corn in the Crib,	231
Liniment for Sprains, etc.,	223	To measure the Contents of Cisterns,	231
Lotion for Mange,	223	Table of Legal Weights,	233
Measurement of Cribs,	231	Uterine Stimulant,	223
Measurement of Cisterns,	231-232		
Measure, Long	232	Wounds in Horses,	220
Measure, Surveyor's	232	Worm Ball for Horses,	220
Mile,	232-233	Weights and Measures,	233-234

MEDICAL DEPARTMENT.

GUIDE TO HEALTH.

WITHOUT health man enjoys but little beneath the sun; and in order that he may have health, so that he can enjoy the life that now is, he must obey the laws of his being. "Cause and effect govern all nature, her pains and pleasures included;" and if we wish to escape her pains and enjoy her pleasures, we *must* obey her laws, for they are "*fixed, wise, and merciful.*" To this end we should give some heed to the following advice:

EATING.—We should eat at *regular hours as much* HEALTHY DIGESTABLE FOOD *as the waste of the system requires*, AND NO MORE. Let breakfast and dinner be the heartiest meals. Never eat heavy or late suppers, nor eat in a hurry, or when you are very tired, or under great mental excitement. Fresh bread and hot biscuits are decidedly unhealthy for all, and especially for dyspeptics; but special rules of diet are of little use here. (See Tables of Food, etc.) In this country there are more children fed to death than starved to death; and we all should eat less and breathe more than we do, in order to have good health.

DRINKING.—We should never drink water within half an hour or an hour of eating, and never drink ice-water, or water nearly as cold as ice, at our meals; and never drink more than half a pint of ordinary cold water at a meal, and but little or no water for two or three hours after meals.

Never drink more than two small cups of tea or coffee at a meal, and they should not be very strong or *hot;* moderately warm is best, and good black tea is the healthiest drink.

We should never drink cold water when the body is overheated by exercise. If we are very warm and thirsty, and wish to drink, we should first wash out our mouths three or four times with cold water, which will so cool the system and slake our thirst that we will then desire to drink but little, and that we can do if we drink *slowly.*

On a hot day we should avoid, as much as possible, drinking cold water in the forenoon; if we drink but little in the forenoon, we will feel the better for having done so the rest of the day.

SLEEPING.—We should go to bed early, and get up early, and have regular hours, as far as possible, for so doing. The young require more sleep than the old; and those who labor, either mentally or physically, require more sleep than those of like age who labor not at all.

From seven to eight hours of sleep are required in every twenty-four hours (and night is the proper time therefor), by those who perform a great amount of physical or mental labor, and from six to seven hours for those who spend their time in idleness. Children require from ten to twelve hours of sleep daily.

We should never go to sleep if the body is hot, even in hot weather, without some covering over us, and never go to sleep with a draught or current of air striking the body.

SLEEPING-ROOMS.—Our bed-rooms should be dry, large, and airy, and be thoroughly ventilated through the day. We should never sleep in a damp bed or in damp bed-clothes, nor allow any damp clothing to remain over night in the bed-room.

All bed-clothing—and indeed all clothing—should be thoroughly dried on coming from the iron; and if not needed for present use, but packed away or put on a spare bed, they should again be dried before using, for they gather dampness, even in dry weather, while thus packed away.

On going to bed, our feet should be dry and warm; but we should have no fire in the bed-room, unless we have sickness, or very cold or damp weather.

DAMP HOUSES.—It is not uncommon for people to sit or sleep in damp houses, yet nothing is much more unhealthy. Besides the dampness caused by wet or damp weather, a great many people have their houses, or the rooms usually occupied, mopped every day or two. This makes them very unhealthy to sit or sleep in, and is the cause of a great deal of unnecessary sickness. Some think that this wetting the floors can do no harm, and especially if there is a good fire in the room at the time; but the fire rather makes the matter worse. When it is absolutely necessary to have rooms mopped or scrubbed, let the family, and especially the children, occupy other rooms until those have become thoroughly dry. We should never sit in a room on a cold, damp day without a fire, if we feel chilly without one.

Some are in the habit of taking down their stoves early in the spring—in April,—and leave their houses thus exposed to all the cold spring rains. This is decidedly very improper. In this climate we should let our stoves remain up until about the first of June, so that we can have fires in our sitting-rooms during the cold or wet spells through the spring season; and it would be healthier to let them remain up all summer, and have fires made in them at least once a month through the summer season.

VISITING SICK ROOMS.—We should never enter a sick room with an empty stomach, or while in a profuse perspiration, if the disease is of a contagious nature; nor stand or sit where a draught carries the air from the bed in the direction of ourselves, nor between the bed and the fire, nor swallow our spittle while in the room, nor eat or drink anything after leaving the room until we have washed out our mouths thoroughly.

EXERCISING.—We all, and especially the young, require a good deal of physical exercise daily, in the open air, in order to enjoy good health. We must have exercise by working, walking, riding on horseback, or in some other way, or we cannot enjoy that health so essential to man for his usefulness and happiness in life.

If we stand or walk, we should carry the body *erect*, and never sit in a cramped, stooping position while reading, writing, sewing, or at any other employment. Our breathing should be deep and full, in order to expand the chest and strengthen the lungs.

BATHING.—We should bathe daily in the summer season, and once or twice a week in the winter, in cold, tepid, or warm water.

If we bathe in cold water, we should do so in the morning, on rising from the bed, wetting the head and face first; or at least we should never bathe in cold water when the powers of the system are exhausted, either by labor or exercise, and never in such cold water as to leave a chilly sensation after our clothes are on. We should wipe dryly, by using a coarse towel *briskly*. The weak and sickly should begin by using warm or tepid water, and as they gain strength they might lower the temperature of the water until it becomes cold. But they should never use water so cold as to make them feel chilly after leaving the bath; such would do more harm than good. The temperature of the water should at all times be regulated by the strength and vitality of the system.

TAKING COLD.—A cold is not necessarily the result of a high or low temperature, but the result of a sudden closing of the pores of the skin, which is produced by cold air upon the skin when it is warm, or the sudden lowering of the temperature of the body in some way. This may be produced in many ways. When the body is warmer than usual, either from exercise or sitting in a warm room, there is a sudden exposure to a still, cold air, or to a raw, damp atmosphere, or to a draught, whether at an open window or door, or street corner, a cold is a certain and an inevitable result.

If the body is over-heated by walking or other exercise, the moment we stop motion we should throw a coat or shawl over our shoulders; and this precaution is absolutely necessary in the warmest weather, and especially if there is the slightest air stirring.

In going into a cold atmosphere, we should keep the mouth closed, and walk with a rapidity sufficient to keep off a feeling of chilliness.

We should protect the body by proper clothing, and never sacrifice health and comfort to appearance. Let us keep the head cool and the feet dry and warm, and avoid draughts of air and a sudden closing of the pores of the skin, and we will be proof against cold and its results.

CLOTHING.—We should at all times protect the body by proper clothing, and never sacrifice our health and comfort, or that of our children, to appearance.

We should at all times wear clothing enough to keep off a chilly sensation; and when the labors of the day are over we should put on a coat, lest there be too sudden lowering of the temperature of the body, and colds or other diseases be the result.

If our clothing has become wet from rain or other causes, we should keep in constant exercise as long as we have them on, and as soon as possible we should exchange them for dry ones.

In cold or wet weather our shoes should be warm and have thick soles, in order to keep the feet warm and dry.

2

From a violation of this rule thousands have met a premature death. Pride is killing more than accidents; and those who sacrifice their own health, or the health of their children to appearance, make an unfortunate exchange. Tight lacing is hurrying thousands to consumptive graves.

While it is true that the young and healthy do not require as much clothing as the old and infirm, yet there are many, and especially little girls and infants, whose health is injured for life, and that often cut short, for the want of *some* clothing on their *arms* and *shoulders*.

The practice of making girls' dresses to rest on the arms or points of the shoulders, is not only disgusting to common sense, but equally uncomfortable and unhealthy for them.

On the danger of exposing the limbs, an able writer says, "A distinguished physician, who died some years since in Paris, declared, 'I believe that during the twenty-six years I have practiced my profession in this city, twenty thousand children have been carried to the cemeteries — a sacrifice to the absurd custom of exposing their arms naked!' I have often thought if a mother was anxious to show the soft, white skin of her baby, and would cut a round hole in the little thing's dress, just over the heart, and then carry it about for observation by the company, it would do very little harm; but to expose the baby's arms — members so far removed from the heart, and with such feeble circulation at best — is a most pernicious practice. Put the bulb of a thermometer in a baby's mouth, and the mercury rises to ninety degrees. Now carry the same to its little hand, and, if the arms be bare and the evening cool, the mercury will sink to fifty degrees. Of course all the blood which flows through those arms must fall between thirty and forty degrees below the temperature of the heart. Need I say when these currents of blood flow back into the chest, the child's general vitality must be more or less compromised? And need I add that we ought not to be surprised at its frequent recurring affections of the tongue, throat, or stomach? I have seen more than one child with habitual cough and hoarseness, or choking with mucous, entirely or permanently relieved by simply keeping its arms and hands warm. Every observing and progressive physician has daily opportunity to witness the same cure."

SCHOOLING.—This thing of sending children who are only four or five years old to school is decidedly wrong. Should they learn a little during one term, they would forget it in a few months, unless they were confined to their books almost all the time, and that would be very injurious to their health, for youth is the time in which to lay up a good constitution for old age; and that cannot be done without a good deal of daily exercise in the open air, with but little study.

I have known parents send their children to school, more to get rid of their noise and the trouble of attending them, than for the purpose of their education! Children should not be sent to school for one quarter, under seven or eight years of age, unless they are allowed more exercise than they now receive, and with less required at their hands. This early confinement and hard study not only produces blue veins and pale faces, but in many instances sickness and premature death. But this evil does not stop here. The young student, in the pursuit of knowledge, carries it into his library, and night after night trims the midnight lamp, in order that he may be thoroughly schooled in that art or science he has chosen for a profession; and by the time he has mastered his studies, his constitution has become so impaired that he is unable to pursue his profession with that degree of application he otherwise could.

But do not suppose, kind reader, that from what I have said I am opposed to a liberal education and a thorough mastery of your profession,—*far from it;* but I know that without health your profession would not have half the charms, nor be half as valuable to you, as it would if you enjoyed good health; *and that you cannot have unless you take exercise, almost daily, in the open air.*

I have known persons with but a limited education, and who knew but little of their profession, accomplish more than good scholars, who were far their superiors in the same profession, and all from the fact, that the former were in the enjoyment of good health, and able to put all their knowledge into practice.

You may now be in the possession of a good constitution, able to accomplish Herculean tasks, and think that hard study and close confinement could not injure your health, but in this you are mistaken. Although the effect of such a course might not be perceptible for some time, yet I think those who disregard this advice will within a few years have the misfortune to experience the truth of these remarks.

PRESENCE OF MIND.—There is no branch of practical education of greater importance than teaching presence of mind. Disasters which occur are greatly increased by the fright and perturbation which are generally manifested on such occasions.

Fright and confusion often result directly from conscious ignorance, and a feeling of inability to help one's self. Hence it is of the utmost importance to fix clearly and indelibly in the mind at all times what course should be pursued when accidents occur. The following are a few rules to be observed in cases of emergency or accident:

If a house takes fire, instantly endeavor to keep all the doors shut. Currents of air and of flame cannot pass through, and it will burn much more slowly; the furniture may be saved, and perhaps the conflagration be so retarded until it may be extinguished.

If the fire is in the chimney, throw into the fire, in the grate or fire-place, or in the stove, a quantity of sulphur, and continue to burn sulphur until the fire in the chimney is put out.

If the lower story is in flames, and inmates are above, the first thing is to direct the attention to loosening a bed cord, or tying bed-clothes together, which, fastened to the bedstead, will admit a safe descent.

If horses become frightened and run, in all cases *keep your seat*, unless they stop, so that you may jump out safely. Always avoid the extreme folly of seizing the reins from the driver.

If the harness break while ascending a hill in a wagon, instantly turn the horses' heads from the bank or precipice, if there be any. This will cause the wheels, in backing, to turn to the same side, and prevent falling or running off. The same precaution is to be observed if a balky horse should commence backing; and if you fill his mouth with dirt or gravel from the road, it will usually have the effect to make him go ahead.

To save horses from a rapidly burning barn, throw a harness or saddle (that to which they have been accustomed) on their backs, and usually they will come out without further trouble; but if they do not, they must be instantly blindfolded.

If from any wound the blood spirts out in jets, instead of a steady stream, you will die in a few minutes unless it is remedied, because an artery has been divided, and that takes the blood direct from the fountain of life. To stop this instantly tie a knot in the middle of a pocket-handkerchief (or, if one is not to be had, use a suspender), then tie the handkerchief loosely around the part cut, *between* the cut and the *body*, placing the knot about a couple of inches from the wound; put in a short stick through the bandage, and twist until the blood stops running, and so keep it until the surgeon arrives.

If the blood flows in a slow, regular stream, a vein has been pierced, and the handkerchief must be bound on the other side of the wound from the heart—that is, *below* the heart.

To allay inflammation, or prevent lock-jaw or mortification in such cases, the following treatment is said to be good: Saturate small pieces of rags of woolen material (raveling of hose or flannel) with grease, (lard or sweet oil) which place upon ignited wood, coal, or charcoal, in an iron kettle, so that they smoke without blazing. Hold the wound over the smoke, if convenient, covering the whole with a blanket, to condense the smoke upon the wound. The kettle should be in or near the chimney, or the windows open at the top, to prevent the deadly effect of inhaling the smoke.

How to Act when the Clothes take Fire.—The "Scientific American" says three persons out of four would rush right up to the burning individual, and begin to paw with their hands, without any definite aims. It is useless to tell the victim to do this or that, or call for water. In fact, it is generally best not to say a word, but seize a blanket from a bed, or a cloak, or any woolen material at hand, hold the corners as far apart as you can, stretch them out higher than your head, and running boldly to the person, make a motion of clasping in the arms, mostly about the shoulders. This instantly smothers the fire and saves the face. At the same time throw the unfortunate person on the floor. This is an additional safety to the face and breath, and any remnant of flame can be put out more leisurely. The next instant immerse the burnt part in cold water, and all the pain will cease with the rapidity of lightning. Next get some common flour, and cover the burnt parts thickly with it, put the patient to bed, and do all that is possible to soothe until the physician arrives. Let the flour remain until it falls off, and a beautiful new skin will be found. Unless the burns are deep, no other application is needed. The dry flour for burns is the most admirable remedy ever proposed, and the information ought to be imparted to all. The principle of its action is that, like the water, it causes instant and perfect relief from pain, by totally excluding the air from the injured parts. Spanish whiting and cold water, of a mushy consistency, are preferred by some. Spread on the flour until no more will stick, and cover with cotton batting.

SAVING DROWNING PERSONS.—Persons who swim and frequent the water ought to have some good plan of procedure impressed on their minds, so that in case of accident it might be of service. A recent treatise on the art of swimming says of the rescue of drowning persons:

"If you have any distance to swim, the wisest plan is to undress, which can be done in a few seconds. You have then more freedom of limb, and can rush through the water with speed and alacrity; and if the drowning person should succeed in clutching you, your chances of freeing yourself, being naked, are innumerable, compared with what they would have been had you been hampered with your wet clothing. When you approach the drowning person, watch diligently for an opportunity, and seize him by the back of the arm, below the shoulder. You will, in this position, be enabled to keep him at arm's length before you, and exercise the most perfect control over his and your own movements. His face being from you, the temptation to grapple with you is removed, and you have more facility to make to the shore or the most convenient place of landing. Never seize a drowning person by the hair of the head. There is great danger to be apprehended in so doing; for, as the arms are at liberty, you are liable to be caught in a death grip at any moment."

For apparently drowned persons, see Recipes.

POISONS AND THEIR ANTIDOTES.—Doctor William Buchan, in his "Domestic Medicine," says: "Every person ought in some measure to be acquainted with the nature and cure of poisons. They are generally taken unawares, and their effects are often so sudden and violent as not to admit of delay, or allow time to procure the assistance of physicians.

Happily, indeed, no great degree of medical knowledge is here necessary; the remedies for most poisons being generally at hand, or easily obtained, and nothing but common prudence needful in the application of them. The vulgar notion, that every poison is cured by some counter-poison, as a specific, has done much hurt. People believe they can do nothing for the patient unless they know the particular antidote to that kind of poison which he has taken; whereas, the cure of all poisons taken into the stomach, without exception, depends chiefly on discharging them as soon as possible. There is no case wherein the indications of cure are more obvious. Poison is seldom long in the stomach before it occasions sickness, with an inclination to vomit. This shows plainly what ought to be done. Indeed, common sense dictates to every one, that if anything has been taken into the stomach which endangers life, it ought immediately to be discharged. Were this duty regarded, the danger arising from poisons might generally be avoided. The method of prevention is obvious, and the means are in the hands of every one."

Poisons either belong to the animal, vegetable, or mineral kingdoms.

Mineral poisons are commonly of an acrid or corrosive quality, as arsenic, cobalt, the corrosive sublimate of mercury, etc.

Those of the vegetable kind are generally of a narcotic or stupefactive quality, as poppy, hemlock, henbane, berries of the deadly nightshade, etc.

Arsenic is the most common of the mineral poisons; and as the whole of them are pretty similar, both in their effects and method of cure, what is said with respect to it will be applicable to every other species of corrosive poison.

When a person has taken arsenic, he soon perceives a burning heat, and a violent pricking pain in his stomach and bowels, with an intolerable thirst, and an inclination to vomit. The tongue and throat feel rough and dry; and if proper means be not soon administered, the patient is seized with great anxiety, hiccoughing, faintings, and coldness of the extremities.

On the first appearance of these symptoms, give immediately from a teaspoonful to a tablespoonful (according to age) of ground mustard, and the same quantity of common salt, in half a pint or less of water, warm or cold, warm being the best, which usually makes the patient vomit almost immediately, and which may be repeated every five or ten minutes. If no mustard is on hand, give large quantities of new milk and salad oil, or oil and warm water, or fresh butter melted and mixed with warm water; or if none of these are at hand, three or four grains of tobacco (a small quid) will usually operate as a ready emetic. Some of these, or other emetics, are to be taken as long as the inclination to vomit continues, or until the stomach has become entirely empty.

Lest there be any remnant of the poison, however small, or its effects, left in the stomach, let the patient take a dose of the best antidote on hand; and if nothing better at hand, give the white of an egg, or a cup of strong coffee.

Should there be any delay, from any cause, in administering an emetic, give immediately a dose of the best antidote you have at hand for the poison taken. It is sometimes best to give an antidote first, and then an emetic, and after thoroughly vomiting, and the stomach has become quieted, give another portion of the antidote. But above all things give an emetic as soon as possible, as life may depend on a prompt and thorough cleansing out of the stomach

POISONS, MINERAL AND VEGETABLE.—*Arsenic* (*Ratsbane*).—Give an emetic; then the white (albumen) of eggs, lime water, chalk and water, or calcined magnesia, and the preparations of iron, particularly hydrate.

Alcohol.—Give an emetic; then dash cold water on the head, and give ammonia (spirits of hartshorn).

Belladonna (Henbane).—Give an emetic; then plenty of vinegar and water, or lemonade.

Corrosive Sublimate.—Give a strong solution of pearlash or salæratus, if at hand; if not at hand, give the white of eggs, or wheat flour and water freely, and then give an emetic.

Creosote.—Give the white of eggs, and then an emetic.

Laudanum.—Same as opium, which see.

Mushrooms, when poisonous.—Give an emetic; then vinegar and water, or either, freely.

Muriatic Acid.—Give an emetic; then calcined magnesia, or soda and water, or salæratus and water, or any alkali.

Nitric Acid (Aqua fortis).—Same as muriatic acid.

Nitrate of Potash (Nitre—Saltpetre).—Give an emetic; then sweet oil, or flaxseed tea, or milk and water, freely.

Nitrate of Silver (Lunar Caustic).—Give a strong solution of common salt, and then an emetic.

Nux Vomica.—Give an emetic; then brandy.

Opium.—Give an emetic; then strong coffee and acid drinks, and dash cold water on the head.

Oxalic Acid.—Give an emetic; then chalk or magnesia, or soap and water, freely; or these first, and then an emetic.

Prussic Acid.—Give an emetic; or first give soda and water, or salæratus and water, or any alkali, and then give an emetic, and pour acetate of potash and common salt, dissolved in water, on the head and spine.

Sulphuric Acid.—Same as muriatic acid.

Strychnine.—Give an emetic; then oil, lard, or fresh butter, and gum camphor in almond mixture, or pounded and mixed with warm water.

Tobacco.—Give an emetic; then astringent teas, and then stimulants.

FOR THE FOLLOWING POISONS EMETICS ARE NOT RECOMMENDED:

Snake Bite.—If the bite is on a limb, instantly tie a cord tightly above the part bitten, and then apply a cupping glass on the bite, and bathe it with spirits of hartshorn. Take a dose of sweet oil, drink spirits freely, and take a tablespoonful of the juice of the tops of green hoarhound three times a day.

Alkalies.—Give ginger.

Ammonia.—Give lemon juice or ginger; then milk and water, or flaxseed tea.

Carbonic Gas.—Remove the patient to the open air, and dash cold water on the head and body; hold hartshorn to the nose, and at the same time rub the chest briskly.

Lead, White Lead, or Sugar of Lead.—Give alum, castor oil, and Epsom salts.

Tartar Emetic.—Give tea made out of galls, Peruvian bark, or white oak bark, freely.

Verdigris.—Give the white of eggs and water, freely.

White Vitriol.—Give milk and water, freely.

HYDROPHOBIA.—*First Symptoms.*—One of the earliest signs of madness in dogs, and one which should always arouse attention on the part of those in charge of dogs, is a sullenness combined with fidgetiness. When it means *rabies* (madness), the dog retires to his bed for several hours, and may be seen curled up, his face buried between his paws and breast. He shows no disposition to bite, and will answer to the call, but he answers slowly and sullenly. After a while he becomes restless, seeking out new resting-places, and never satisfied long with one. He then returns to his bed, but continually shifts his posture. He rises up and lies down again, settles his body in a variety of postures, disposes his bed with his paws, shaking it in his mouth, bringing it to a heap, on which he carefully lays his chest, and then rises up and bundles it all out of his kennel. If at liberty, he will seem to imagine something lost, and will eagerly search around with strange violence and indecision. That dog should be watched. If he begins to gaze strangely about him, as he lies in bed, and if his countenance is clouded and suspicious, we may be certain that madness is coming on.

Cure for Hydrophobia.—James Williams, M. D., writes to the "Philadelphia Inquirer" an interesting letter on the treatment of persons bitten by mad dogs. He himself was bitten, and was cured by the mode prescribed in the following extract from his letter:

"When any person is bitten, with a sharp knife cut away a small portion of the flesh surrounding the wound, and cauterize the part freely with lunar caustic, and repeat the application for two or three days; dress the wound each time with a little simple cerate or fresh lard.

"Take of the root of elicampane (*cenula campana*) one ounce and a half, cut it fine or bruise it, then boil it in one pint of new milk, down to a half pint; give this quantity for three mornings, fasting; do not let the patient eat any thing until about four o'clock in the afternoon; let him be kept quiet, and as confident and cheerful as possible. It is confidently asserted by those who have used the elicampane and milk alone, that it will have the desired effect if taken within twenty-four hours after the accident.

"I would suggest, sir, that some suitable person in every rural district be provided with the proper means for relief, to be ready for use at any time when applied for. Such is my confidence in the above mode of treatment that I would unhesitatingly pursue it to the exclusion of every known or supposed remedy whatever."

To show the effects of early local application, Mr. Williams mentions that an English groom and a boy were bitten at the same time, and by the same dog. Every thing was done for the boy that love or money could suggest, but he died in horrible convulsions. The groom was overlooked in the confusion; and he, thinking lightly of the matter, merely washed his wound with water and strong country or home-made soap, and entirely recovered, never experiencing any ill effects from the bite.

Another Cure.—A Saxon forester, by name Gastell, now of the venerable age of eighty-two, unwilling to take to the grave with him a secret of such importance to mankind, has made public in the "Leipsic Journal" the means which he had used for fifty years, and wherewith, he affirms, he has rescued many fellow beings and cattle from the fearful death of hydrophobia. Take immediately warm vinegar, or tepid water, wash the wound therewith, and then dry it; then pour upon the wound a few drops of *muriatic* acid, because mineral acids destroy the poison of the saliva, by which means the evil effect of the latter is neutralized.

Cure for Hydrophobia.—Doctor Buisson (says the "Salut Public" of Lyons) claims to have discovered a remedy for this terrible disease, and to have applied it with complete success in many cases. In attending a female patient in the last stage of canine rabies, the doctor imprudently wiped his hands with a handkerchief impregnated with her saliva. There happened to be a slight abrasion on the index finger of the left hand, and confident in his own curative system, the doctor merely washed the part with water. However, he was fully aware of the imprudence he had committed, and gives the following account of the matter afterwards:

"Believing that the malady would not declare itself until the fortieth day, having numerous patients to visit, I put off from day to day the application of my remedy—that is to say, vapor baths. The ninth day, being in my cabinet, I felt all at once a pain in the throat, and a still greater one in the eyes. My body seemed so light that I felt as if I could jump to a prodigious height, or that if I threw myself out of the window I could sustain myself in the air. My hair was so sensitive that I appeared able to count each separately without looking at it. Saliva kept continually forming in my mouth. Any movement of air inflicted great pain on me, and I was obliged to avoid the sight of brilliant objects. I had a continual desire to bite—not human beings, but animals, and all that was near me. I drank with difficulty, and I remarked that the sight of water distressed me more than the pain in the throat. I believed that by shutting the eyes any one suffering under hydrophobia can always drink. The fit came on every five minutes, and I then felt the pains start from my index finger, and run up the nerves to the shoulder. In this state, thinking my course was preservative, and not curative, I took a vapor bath—not with the intention of cure, but of suffocating myself. When the bath was at a heat of fifty-two degrees centigrade all the symptoms disappeared, as if by magic, and since then I have never felt anything more of them. I have attended more than eighty persons bitten by mad animals, and I have not lost a single case."

When a person has been bitten by a mad dog, he must for seven successive days take a vapor bath *à la Russe*, as it is called, of fifty-nine to sixty-three degrees. This is the preventive remedy. When the disease is declared it only requires one vapor bath, rapidly increased to thirty-seven degrees centigrade, then slowly to sixty-three degrees. The patient must strictly confine himself to his chamber until the cure is complete.

Doctor Buisson mentions several other curious facts. An American had been bitten by a rattlesnake, about eight leagues away from home. Wishing to die in the bosom of his family, he ran the greater part of his way home, and going to bed, perspired profusely, and the wound healed as any simple cut. The bite of the tarantula is cured by the excess of dancing, the free perspiration dissipating the virus. If a young child be vaccinated, and then be made to take a vapor bath, the vaccine does not take.

Another Cure for Hydrophobia.—A writer in the "National Intelligencer" says that spirits of hartshorn is a certain remedy for the bite of a mad dog. The wounds, he adds, should be constantly bathed with it, and three or four doses, diluted, taken inwardly during the day. The hartshorn decomposes chemically the virus insinuated into the wound, and immediately alters and destroys its deleteriousness. The writer, who resided in Brazil for some time, first tried it for the bite of a scorpion, and found that it removed pain and inflammation almost instantly. Subsequently he tried it for the bite of a rattlesnake, with similar success. At the suggestion of the writer, an old friend and physician in England tried it in cases of hydrophobia, and always with success.

Another Cure.—Mix one pound of common salt in a quart of water, and then bathe with and squeeze the wound with it one hour, then bind a little more salt on the wound for twelve hours. The author of this receipt was bitten six times by mad dogs, always cured himself with the above mixture, and offered to suffer himself to be bitten by any mad dog, in order to convince mankind that what he offered was a real truth, to which numbers could testify.

Another Cure.—A correspondent of the "Providence Journal" sends that paper the following receipt, as a remedy for hydrophobia: Eat the green shoots of asparagus raw; sleep and perspiration will be induced, and the disease can thus be cured in any stage of canine madness. A man in Athens, Greece, was cured by this remedy after the paroxysms had commenced.

Singular Case of Hydrophobia—A Man becomes Mad almost daily for nearly Twenty Years—Discovery of a Preventive of this dreadful Disease.—The following is taken from the "Cincinnati Inquirer:"

"A remarkable case of hydrophobia, that demon which yearly frightens all mothers, and with its horrible spasms sends to the land of the hereafter almost all who become infected by it, came to our knowledge a few days ago, and may be of interest, inasmuch as it suggests the possibility of a cure.

"Almost twenty years ago, a man named Clarke, who resides in Jamestown, Kentucky, a little town about three miles from this city, was bitten by a dog which proved to be rabid, and in a short time afterwards was taken with the most violent symptoms of that terrible disease.

"The malady, which, as is well known, sometimes exists in the system for a number of years (one or two cases are known of twenty years' standing) before it makes its appearance, did not in this instance prove immediately fatal, and by the exertions of his physicians and friends the spasms were for a short time allayed, and the patient regained a good degree of health.

"A few days, however, only elapsed when he again perceived a recurrence of his disposition to bite and snap, together with the hatred of water, and a spasmodic contraction of the throat, by which the disease is characterized, and he had a severe and much more violent attack than the first, during which even the physicians who attended him gave up the case as hopeless, and left him to die. By some means, however, this fact came to the knowledge of some person who had heard (from what source we cannot tell) that a medicine known to botanic physicians as the 'third preparation,' had been found beneficial in such cases, and was recommended to try it. His friends, who had no hope that he could be saved, at first thought it would be kindness to allow death to end his misery, and refused to make any attempt to further prolong an existence which, if preserved, could only keep all by whom he was surrounded in constant fear of being contaminated, and in danger of violence from their parent and friend.

"After much deliberation they at length concluded to try the experiment, and procured some of the medicine we have named, which is composed of capsicum and the tinctures of lobelia and myrrh, and making it very strong of the tinctures, gave it to him in sufficient dose to produce a thorough emesis. He threw from his stomach a large quantity of frothy mucous, and from that moment the spasms ceased, and there was also a relief from the other more prominent symptoms of the disease. He gradually grew better as this treatment was continued, and at length became able to attend to his duties, which he has done with but slight intermissions ever since. Almost every day, although we believe he has not been at any time confined to his bed, there has been a recurrence of the disease, which, however, has been promptly checked by the same means which wrought such a miraculous change in the first instance. He now attends to his business daily, and when this contraction of the throat makes its appearance he doses himself largely with the preparation, which he keeps constantly about him, and immediately upon this discharge from the stomach becomes well."

Cure for Hydrophobia.—A correspondent of the "St. Louis Republican," Mr. J. A. Hubbard, who had in early youth, together with his brother, been bitten by a mad dog, states that both were cured by drinking a strong decoction made from the bark of the root of the black ash, which is a well-known cure for the bite of a rattlesnake, drinking a wine glass of it three times a day for eight days. This is a very simple remedy, and should at least have a trial. He gives the following as the mode of preparing it: Take the root of the common upland ash, generally called black ash, peel off the bark, and boil it to a strong decoction, and of this drink freely.

Hydrophobia. — *Remarkable Cure by applying the Mad Stone.* We extract the following from the " Utica (Indiana) Ledger:"

"Last week we gave notice of a Mr. Burnett (whom by mistake we called Barrett) being bitten by a supposed mad dog, and of his going to Terre Haute, to apply the 'mad stone.' From Mr. Henry Schlosser we have the particulars of the result. When Mr. Burnett was bitten, he hardly thought the dog mad; but Mr. Schlosser, on whose farm Burnett was a tenant, advised him to be prudent, and lose no time in trying the mad stone. Burnett was bitten on Sunday, and on Tuesday started for Terre Haute. On his arrival the arm and hand were swollen almost to bursting, and the wound was very painful, all four of the dog's tusks having penetrated his wrist. Immediately upon his arrival he repaired to the house of Mrs. Taylor, where the mad stone was applied. The stone adhered to the wound with wonderful tenacity, remaining for twenty-four hours. It was then applied again, on another part of the arm, where it adhered several hours. A third time it was applied to another part of the arm, but it would not adhere, and Mrs. Taylor informed her patient that he was cured—the fact of refusing to adhere being evidence that all the poison was extracted. During the operation a green fluid was absorbed by the stone, and trickled in drops from the wound, and the patient could feel the pain leaving his arm. When the stone became charged with this poisonous matter, it was soaked in warm water for a time, when it would again adhere with its original tenacity, and continued to do so until the greenish matter ceased to appear, when it dropped off.

"Mrs. Taylor pronounced the case of Mr. Burnett as bad, or the worst one she had ever seen; but she very confidently assured Mr. Burnett that after the operation of the stone he need have no farther fears of hydrophobia.

"We have heard often of this mad stone, and, being incredulous, took the pains to make some inquiries concerning it. The stone has been in the possession of the Taylor family for a great many years, and has been tried a great many times on persons bitten by mad animals. It has never failed to cure in a single instance, we are told. Twenty years ago Lewis Toms, of this place, who is now living to attest the statement, was bitten by a mad dog, and was cured by the application of this wonderful stone. We did not learn where this wonderful curative was originally obtained. It is porous, and in appearance much resembles a piece of lava, in common use among painters. Mr. Toms, recollecting its good effect on himself, joined Mr. Schlosser in advising Burnett to try it, and he came back relieved of his intense pain, and feeling that his life was saved by the mad stone. Mrs. Taylor's charges were very reasonable, being fifteen dollars for the operation, and boarding the patient during the time occupied in curing him."

Mad Stone.—Mrs. McNeil, better known as "Sister Elizabeth," formerly of the Catholic school in this city, writes from Valparaiso, says the "Lafayette (Indiana) Courier," stating that a lady of that place has a mad stone which has effectually cured more than fifty persons who had been bitten by mad dogs. The stone acts as a leech, and absorbs all the poison when applied to the wound. The lady in question has refused one thousand dollars for it. She applies it free on personal application at her home in Valparaiso, and in extreme cases will visit distant cities upon the guarantee of her expenses.

A lady residing in Terre Haute, says the "Journal" of that city, is also in possession of a valuable mad stone, which has effected many remarkable cures in cases of dog and snake bites. Hundreds of persons, some from a great distance, have tested its efficiency in such cases.

The Mad Stone.—On this interesting subject the "Henry (Marshall county, Illinois,) Courier" says:

"We stated in our last that a Mr. Mallory and Mr. Ward, of this city, had been bitten by a mad dog, and had gone to try the virtue of a 'mad stone' in the central part of this State. The parties returned home on Monday evening last, having applied the stone, and have every reason to believe that they are cured. Mr. Mallory called at our office on Tuesday, and gives the following account of the operation.

"He found the mad stone in the possession of Mr. J. P. Evans, of Lincoln, Logan county, and describes it as a small flesh-colored stone, about two inches broad, half an inch thick, and very porous. The stone was first placed in warm water for an hour, and applied to the flesh wound, when it adhered firmly for several hours, all the time apparently drawing, with a strong suction, the blood from all parts of the body. After remaining on several hours, the stone, as it became charged with the poison, became of milky whiteness, as did also the flesh immediately about the wound; when all at once it fell off, and being placed in warm milk, emitted a strong, offensive odor, and gradually discharged its contents into the milk, and assumed its natural color again. It was then again applied, with the same results, several times, until finally it would adhere no longer, and the patient was declared cured.

"The parties have all confidence in the treatment, and feel an assurance that they have averted an awful death, particularly as it has since been ascertained that the dog that bit them was really mad, and a mare belonging to Mr. Bickerman, which was bitten the same day, has already been attacked with hydrophobia, and is probably dead by this time.

"Mr. Mallory states that there were several other patients, from various parts of the country, waiting at Mr. Evans', to have the stone applied, and that it had never been known to fail in effecting a cure."

The Mad Stone a Humbug.—"We learn," the "Springfield Register" says, "that Mr. Robert Forsyth, of Lincoln, died last Wednesday, from hydrophobia. Mr. F. was bitten about six weeks ago, by a dog that was supposed to be mad, and he immediately made an application of the famous mad stone belonging to Mr. Evans, but it seems without any benefit, for he died in the greatest agony at the time mentioned."

The Mad Stone.—The "Tazewell Register" says: "One of these stones, supposed to be so efficacious in cases of hydrophobia, is in the possession of Mr. Joshua Houser, who resides about four miles west of Alanta, in McLean county. It has been in the possession of that family for over fifty years, having been given to him by his mother, in Richmond, Virginia. The virtues of this stone have recently been tested by a gentleman residing near Delevan, whose child was severely bitten by a pet dog, which exhibited all the symptoms of hydrophobia. He desires us to call attention to the fact, that Mr. Houser is very reasonable in his charges for applying the stone and attending to sufferers. He is more influenced by a desire to alleviate sufferings than to speculate upon those who are so unfortunate as to need his services. In no case does his charge exceed ten dollars."

"*The Mad Stone.*—The subscriber, living near Holly Springs, Mississippi, owns the other half of the celebrated " Snake Stone," sometimes called the " Mad Stone," (it having been broken in two some forty years ago,) the other half being recently sold in Virginia at the sale of the late Samuel Pointer, of that State, for the sum of six hundred dollars. This stone has been owned by my father and myself for the last forty years, during which time it has been applied to at least FOUR HUNDRED PERSONS who were *bitten by mad dogs*, not one of whom has had hydrophobia! I do not hesitate to believe that the application of it will prevent the development of that fearful disease. It can be found at all times at my house. Price of application, twenty dollars; to be refunded to the friends of any person who may die of hydrophobia after the application of the stone.

"DAVID POINTER."

It is said that strong vinegar will effect a cure, though I would not rely on this alone. If the bite was on a limb, I would instantly apply a ligature tightly above the part bitten; then, with a sharp knife cut away a small portion of the flesh surrounding the wound, and apply a cupping-glass on the wound; then cauterize the part freely with nitrate of silver. This treatment, to be followed up, with others, as suggested in some of the recipes, would, I think, cure the bite of a mad dog.

CHOLERA.—*Treatment in Turkey.*—The American missionaries in Constantinople have been successfully employed in treating cases of cholera and cholerine during the progress of that fearful epidemic in that city. Of some four hundred cases under their care, the mortality did not exceed five per centum of the whole number; and it is remarked that most of them were treated under very unfavorable circumstances, in the almost inconceivably filthy and crowded rooms of the Khans, and in several cases the patient was almost in a dying state when first seen. The general course of treatment adopted has been as follows: In the first stages of the disease, before the collapsed state, it has been their sole object to control the purging and vomiting. In every case where these symptoms have been overcome, recovery has followed. The application of strong mustard plasters to the abdomen has never been neglected in any case; and where the disease has made much progress, the mustard has been mixed with vinegar instead of water. To check the diarrhœa the Reverend Doctor Hamlin's cholera mixture has been used, namely, a mixture of equal parts of laudanum, tincture of rhubarb, and spirits of camphor. The initial dose is thirty drops for an adult; but in cases of real cholera the first dose may be sixty drops, to be increased, say ten drops after each discharge, up to one hundred, or even one hundred and twenty-five drops, and to be continued in small and decreasing doses every three or four hours for twenty-four hours after the cessation of the discharges. It is to be remembered that no danger exists from the medicine until it controls the discharge.

In cases of vomiting they have used a mixture given to Doctor Hamlin, by a physician of great experience in the treatment of cholera, consisting of equal parts of laudanum, tincture of capsicum, tincture of ginger, and tincture of cardamom seeds. Initial doses from thirty to forty drops, but in severe and obstinate cases of vomiting a teaspoonful may be given, mixed with two teaspoonsful of brandy and a little water. This mixture is also useful in certain cases of diarrhœa, and may be given in the same dose as the first mixture. In certain cases, where these mixtures have failed to stop the evacuations, they have been exchanged for prepared chalk (*creta preparata*), say fifteen to twenty grains of chalk, with ten drops of laudanum and ten drops of spirits of camphor, administered after each evacuation, have proved more efficacious than anything else, and they should be tried in every severe case. In the collapsed state there is always coldness of the extremities, often with severe cramps, sinking of the eyes, shriveling, and often discoloration of the face, with a general sinking of the whole system. For persons found in this state there is much less hope. In these cases every effort must be made to keep up the circulation of the blood, with mustard plasters on the abdomen, feet, calves of the legs, and back of the neck. Baths of hot water or hot bricks should surround the patient, and brandy has been used in frequent cases, as a stimulant, with good effect. Ammonia ointment has been used on the spine, in the collapsed state, with the best effect.

When typhoid symptoms follow the subsidence of an attack of cholera, it has been found that in almost every case, with a rice-water diet and the free use of chamomile tea, these symptoms would disappear of themselves in from one to five days. Small doses of sweet spirits of nitre may be given, however, and in some cases small doses of calomel and opium have been found necessary to the restoration of health. When the patient is very weak, quinine or gentium may be given. Intense thirst is common in cholera cases. For this gum arabic water and chamomile tea may be given in small quantities, but the drinking of water should be carefully prevented. Great care should be taken of the diet during and after an attack of cholera. On the first day nothing should be eaten; afterward, rice water, arrow root cooked with water, chamomile tea, and such things, should be chosen.

The above method of treatment is not presented as a substitute for the aid of good physicians, who should always be called at once when they can be found, but simply as a guide to those who may be compelled by circumstances to act by themselves. By the timely use of powerful medicines, cholera may be prevented in ninety-eight cases out of a hundred, where instant attention is given to the first signs of diarrhœa.

How to avoid Cholera.—At a meeting of the New York Health Commissioners, a short time since, Doctor Sayres, of Brooklyn, gave the following general rules for the treatment of cholera patients:

"Great fear and anxiety has a great deal to do in the prostration and spread of disease, and influences the physical condition to such a degree as to make persons more subject to it than they would otherwise be. I think if the people understood the single fact, that the cholera is not necessarily a fatal disease, and that it is always preceded by certain premonitory symptoms, such as lassitude, great languor, debility, and a diarrhœa, and that in this stage of the disease it is nearly always curable, if the proper precautionary measures are taken, it would tend to allay the popular terror. At this stage of the disease it is of the first importance to pay attention to the first symptom, which is diarrhœa. At the very first approach the patient should assume a horizontal posture, and retain it, with the hips higher than the shoulders, and under no circumstance assume the perpendicular, even for a moment. Absolute, positive rest is needed—the body being kept in a warm condition. Any artificial means can be used for that which may be necessary. The main thing is the horizontal position and perfect rest at the very commencement of the disease. If the patient is down stairs when taken, let him stay there or be carried up, but do not let him walk up. If visiting out, let him stay at his friend's house. Keeping this position for forty-eight hours, in the majority of cases the disease will pass away, and the patient, on recovery, need make no very great changes in his mode of life. Of course, if he has bad habits he should reform them, eat and drink rationally, and attend to business as usual, but not overdo himself. In the present state of the atmosphere, when pestilence is abroad, the system is rendered more liable to exhaustion, and he should husband his strength by avoiding violent exercise. With these few precautionary measures, a proper sanitary condition of the city, and a rigid enforcement of quarantine, we may hope to escape any pestilential epidemic."

Ozone and Cholera. — A correspondent of "Galignani's Messenger" states the following curious fact:

"Immediately after a short but violent thunder storm at Ancona, a great improvement in the public health was observed. The number of cases of cholera diminished considerably, and in most cases the patients recovered almost immediately from the prostration and languor which accompany the disease. This goes to prove the correctness of the supposition which has been thrown out before, that the prevalence of cholera is due to the absence of ozone in the atmosphere. Ozone is a gaseous substance, the nature of which is not well understood. It is commonly regarded as a certain modification of oxygen. We only know that it can be produced by electrifying the air, and consequently it always abounds during thunder storms. It would not be amiss to try the experiment of generating it in the chambers of cholera patients. The process is very simple: it is only necessary to pass a succession of electric sparks through the air. It was ascertained a few years ago, by observations made by a physician in Russia, at a time when the cholera was raging among the soldiers, that the disease is always preceded by a lowering of the pulse, even down to forty. If ozone, therefore, as it would appear, acts as a quickener of the vital powers, we can easily see why its presence should have such a remarkable effect in checking the progress of the epidemic."

Effective Mode of Fumigation. — In the "New York World" we find the following description of the manner of fumigating a ship infected by cholera:

"Doctor Sayers has originated a new method of fumigating and disinfecting vessels, which he thinks is thorough and reliable. The old plan of burning tar is not sufficient to destroy the infection of clothes, and he therefore adopts chlorine as a disinfecting agent. It was adopted on the steamer Europe, which was detained for a while, and will be carried out on the Atlanta before she is allowed to come to the city. The plan is as follows:

"Large shallow pans are made of sheet lead, by bending the edges upward, and numbers of them are placed on the floor and elsewhere in the hold, and staterooms (if there be any) and cabin. A layer of manganese is then spread in each, some two or three hundred pounds being used for a single vessel. The hatches and all other openings to the outer air are fixed, ready to be closely battered down in a moment's time; then hydrochloric acid is poured on the manganese, twenty or thirty carboys being necessary. This is done very quickly, as the deadly gas begins to generate immediately, and the operator hurries to the deck, when the hatches are battered down, and he leaves the ship. All trunks, boxes and chests are previously opened, of course, so that the gas can have free access. Chlorine forms in immense quantities, and permeates everything—perfectly neutralizing all infection, killing all insects and animal life, and deodorizing everything impure or diseased. If by any accident a fire should be burning, the gas extinguishes it at once. The man who puts the acid on the manganese must not tarry, of course, or he would be killed by the gas in less than ten minutes. The vessel is thus left alone for some hours, after which the hatches are opened, ventilators put up, the gas blown out, and the ship is as pure as when first built. Of course it is necessary to transfer the passengers to some other vessel, or to land them, while the process is going on. Moreover, before the whole vessel is fumigated each passenger selects a suit of clothes from his wardrobe, tickets it, and these suits are hung up in a special room, which is closed and fumigated by the same process. Then the passengers bathe, and those with beards shave their faces, if they shave at all, and evacuate the bowels, dress rapidly in their purified clothes, and are immediately transferred to another ship, after which the one they leave is fumigated according to the plan detailed above. This plan is claimed to be far superior to any other yet invented."

What will Cure the Cholera.—From the "London Weekly Dispatch" we extract the following interesting article on this subject:

"Doctor Worms, head physician to the military hospital of the Gros Gaillo, in Paris, some time since read a paper to the Academy of Medicine, offering what would appear to be very nearly a specific in cholera cases. In July, 1849, Doctor Worms gave the results of his experience in a letter to the 'Gazette Médicale.' After describing the failure of other remedies, he had recourse to the administration (in the cure of patients already prostrated, but not in the last stage) of *mineral lemonade,* suppressing all other medicament. This lemonade is a preparation of two, three, or at most four grammes of concentrated sulphuric acid, with one thousand grammes of common water and one hundred and fifty of simple syrup. At the time he was unable, through the abatement of the disease, to test his remedy upon patients in its worst stages; but in three more recent visitations he was enabled to extend his experience, and it is summed up in the assertion—'Recently he has witnessed the infallibility of this simple treatment, and he desires to see it generalized.' In confirmed cholera the patient is left in the most complete repose. He is shampooned only in case of cramps. He takes a glass of the lemonade (from five to ten grammes of acid per litre) every half hour; he may do this immediately after vomiting, which is intended to be prompted while diarrhœa is stopped, the vomiting being a sign of amendment. White wines, champagne, and ice may be used freely, but not beer, brandy, or alkaline mineral waters. Within our own circle we receive testimony corroborating the principle of this remedy. A practitioner has resorted to lemon and salt for the last twenty years or more, with essential success. Late intelligence of a serious outbreak of cholera at Epping, fatal in five instances, bids us be on the watch. Nothing more need be said on the value of pure air and cleanliness. But medical science has hitherto been at fault. Every organ of publicity should therefore place before its readers whatever may seem to offer a hope of better success, that those educated to the healing art may in no instance miss a suggestion that may prove fruitful."

Another Cure.—Take three tablespoonsful of castor oil, the same quantity of the best French brandy, and forty drops of laudanum, well mixed together, and let the patient drink it. The body must then be rubbed over with a hot flannel cloth. Should the condition of the patient not improve within one hour, and the nails of the fingers begin to get black, give one tablespoonful of castor oil, one of French brandy, and ten drops of laudanum. This generally throws the sufferer into a profound sleep, from which he will awaken perfectly well. This treatment is said to be adopted in India.

Another Cure.—To an adult give one grain of opium and four grains of kino, every half hour. To children give one teaspoonful of strong tincture made from cinnamon bark every half hour. Then strip the patient and throw cold water over the whole body, hastily wiping dry, put into blankets to sweat, and give a little well-boiled corn-meal gruel. The cold water coming in contact with the nitrogen gas on the body forms a neutral, the cramps instantly cease, the regular circulation returns, and the patient is said to be cured.

Another Cure.—A tablespoonful of common salt and a teaspoonful of red pepper, mixed in half a pint of hot water, and drank, is said to be very good to arrest the diarrhœa in this disease.

Another Cure for the same.—
Best brandy .... 1 pint.
Cinnamon bark (ground)............... ½ ounce.
Cloves (ground)....................... ½ ounce.
Prickly ash berries or bark............ ½ ounce.
Gum guaiacum......................... ½ ounce.
Gum camphor......................... ¼ ounce.

Dose—from a tea to a tablespoonful every half hour, or oftener, till the disease is arrested.

Another Cure for the Cholera.—The subjoined letter, from a well-known Boston druggist and apothecary, was first published several years ago in a Boston paper. We advise our readers to preserve it:

"The Reverend Doctor Hamilton, of Constantinople, saved hundreds of lives by the following simple preparation, during the terrible raging of cholera in that city a few years since. In no case did the remedy fail when the patient could be reached in season. It is no less effective in cholera morbus and ordinary diarrhœa. A remedy so easily procured and so vitally efficacious should be always at hand. An ordinary phial of it can be had for twenty-five cents, and nobody should be without it over night. The writer of this received the recipe a short time since, and recently having been seriously attacked by cholera morbus, can testify to its almost magical influence in affording relief from excruciating pain. He ardently hopes that every one whose eyes trace these lines will cut this article from the paper, and procure the medicine without delay. Its prompt application will relieve pain, and presumptively save life. Take one part of laudanum, one part of camphorated spirits, two parts of tincture of ginger, and two parts of capsicum. Dose—one teaspoonful in a wine glass of water. If the case is obstinate, repeat the dose in three or four hours."

RULES FOR GENERAL HEALTH.—The following important suggestions, for the benefit of soldiers, are taken from

"Hall's Journal of Health." They are fraught with wisdom, and may be of great value to others:

1. In any ordinary campaign, sickness disables or destroys three times as many as the sword.

2. On a march, from April to November, the entire clothing should be a colored flannel shirt, with a loosely buttoned collar, cotton drawers, woolen pantaloons, shoes and stockings, and a light-colored felt hat, with broad brim, to protect the eyes and face from the glare of the sun and from the rain, and a substantial but not heavy coat when off duty.

3. Sunstroke is most effectually prevented by wearing a silk handkerchief in the crown of the hat.

4. Colored blankets are best, and, if lined with brown drilling, the warmth and durability are doubled, while the protection against dampness from lying upon the ground is almost complete.

5. Never lie or sit down on the grass or bare earth for a moment; rather use your hat — a handkerchief, even, is a great protection. The warmer you are, the greater need for this precaution, as a damp vapor is immediately generated, to be absorbed by the clothing, and to cool you off too rapidly.

6. While marching, or on other active duty, the more thirsty you are, the more essential it is to the safety of life itself, to rinse out the mouth two or three times, and then take a swallow of water at a time, with short intervals. A brave French general, on a forced march, fell dead on the instant by drinking largely of cold water when snow was on the ground. As far as possible avoid drinking during the first part of the day, and but little thirst will be felt during the latter part. The free use of water in the fore part of the day will create great thirst in the latter, which drinking will not satisfy. To relieve the first indications of thirst, place in the mouth a raisin, or the pit of a small date, plum, or cherry, or in their absence a small pebble, which

will moderately excite the saliva and quiet the desire to drink.

7. Abundant sleep is essential to bodily efficiency, and to that alertness of mind which is all important in an engagement; and few things more certainly and more effectually prevent sound sleep than eating heartily after sundown, especially after a heavy march or a desperate battle.

8. Nothing is more certain to secure endurance and capability of long-continued effort than the avoidance of everything as a drink except cold water, not excluding coffee at breakfast. Drink as little as possible even of cold water.

9. After any sort of exhausting effort, a cup of coffee, hot or cold, is an admirable sustainer of the strength until nature begins to recover herself.

10. Never eat heartily just before a great undertaking, because the nervous power is irresistibly drawn to the stomach, to manage the food eaten, thus drawing off that supply which the brain and the muscles so much need.

11. If persons will drink brandy, it is incomparably safer to do so after an effort than before, for it can give only a transient strength, lasting but a few minutes; but as it can never be known how long any given effort is to be kept in continuance, and if longer than the few minutes, the body becomes more feeble than it would have been without the stimulus, it is clear that its use before an effort is always hazardous, and is always unwise

12. Never go to sleep, especially after a great effort, even in hot weather, without some covering over you.

13. Under all circumstances, rather than lie down on the bare ground, lie in the hollow of two logs placed together, or across several small pieces of wood laid side by side; or sit on your hat, leaning against a tree. A nap of ten or fifteen minutes in that position will refresh you more than an hour on the bare earth, with the additional advantage of perfect safety.

14. A cut is less dangerous than a bullet wound, and heals more rapidly.

15. If from any wound the blood spirts out in jets, instead of a steady stream, you will die in a few minutes unless it is remedied, because an artery has been divided, and that takes the blood directly from the fountain of life. To stop this instantly, tie a handkerchief, or other cloth, very loosely between the wound and the heart; put a stick, bayonet, or ramrod between the skin and the handkerchief, and twist it around until the bleeding ceases. Keep it thus until the surgeon arrives.

16. If the blood flows in a slow and regular stream, a vein has been pierced, and the handkerchief must be on the other side of the wound from the heart—that is, below the wound.

17. A bullet through the abdomen (belly or stomach) is more certainly fatal than if aimed at the head or heart, for in the latter case the ball is often glanced off by the bone, or follows it round under the skin; but when it enters the stomach or bowels, from any direction, death is inevitable under all conceivable circumstances, but is scarcely ever instantaneous. Generally the person lives a day or two with perfect clearness of intellect, often not suffering greatly. The practical bearing of this statement, in reference to the great future, is clear.

18. Let the whole beard grow, but not longer than some three inches. This strengthens and thickens its growth, and thus makes a more perfect protection for the lungs against dust, and of the throat against winds and colds in winter, while in summer a greater perspiration of the skin is induced, with an increase of evaporation; hence, greater coolness of the parts on the outside, while the throat is less feverish, thirsty, and dry.

19. Avoid fats and fat meats in summer, and on all warm days.

20. Whenever possible, take a plunge into any lake or

running stream every morning, as soon as you get up. If none is at hand, endeavor to wash the whole body over as soon as you leave your bed; for personal cleanliness acts like a charm against all diseases, always either warding them off altogether, or greatly mitigating their severity and shortening their duration.

21. Keep the hair of the head closely cut, within an inch and a half of the scalp in every part, repeating on the first of each month, and wash the whole scalp plentifully in cold water every morning.

22. Wear woolen stockings and moderately loose shoes, keeping the toe and finger nails always cut close.

23. It is more important to wash the feet well every night, than to wash the face and hands of mornings, because it aids to keep the skin and nails soft, and to prevent corns, chafings, and blisters, all of which greatly interfere with a soldier's duty.

24. The most universally safe position, after stunnings, hurts, and wounds, is that of being placed on the back, the head being elevated three or four inches only—aiding, more than any one thing else can do, to equalize and restore the proper circulation of the blood.

25. The more weary you are after a march or other work, the more easily will you take cold if you remain still after it is over, unless, the moment you cease motion, you throw a coat or blanket over your shoulders. This precaution should be taken in the warmest weather, especially if there is even the slightest air stirring.

26. The greatest physical kindness you can show to a severely wounded comrade is first to place him on his back, and then run with all your might for some water to drink. Not a moment should be lost. If no vessel is at hand, take your hat; if no hat, off with your shirt, wring it out once, tie the arms in a knot, as also at the other end, thus making a bag open at the neck only. A fleet person can convey a bucketful a mile in this way. I have seen a dying man

clutch at a single drop of water from the fingers' end, with the voraciousness of a famished tiger.

27. If wet to the skin by rain, or by swimming rivers, keep in motion till the clothes are dried, and no harm will result.

28. Whenever it is possible, by all means, when you have to use water for cooking or drinking, from ponds or sluggish streams, boil it well, and when cool, shake it or stir it, so that the oxygen of the air shall get to it, which greatly improves it for drinking. This boiling arrests the process of fomentation, which arises from the presence of organic and inorganic impurities, thus tending to prevent cholera and all bowel diseases. If there is no time for boiling, at least strain it through a cloth, even if you have to use a shirt or the leg of a trowsers.

29. Twelve men are hit in battle, dressed in red, where there are only five, dressed in bluish gray—a difference of more than two to one; green, seven; brown, six.

30. Water can be made almost ice-cold in the warmest weather, by closely enveloping a filled canteen or other vessel with a woolen cloth kept plentifully wetted and exposed.

31. While on a march, lie down the moment you halt for a rest; every minute spent in that position refreshes more than five minutes standing or loitering about.

32. A daily evacuation of the bowels is indispensable to bodily health, vigor, and endurance; this is promoted, in many cases, by stirring a teaspoonful of Indian corn-meal in a glass of water, and drinking it on rising in the morning.

33. Loose bowels—namely, acting more than once a day, with a feeling of debility afterward—is the first step toward cholera. The best remedy is instant and perfect quietude of body, eating nothing but boiled rice, with or without boiled milk; in more decided cases a woolen flannel, with two thicknesses in front, should be bound tightly around the abdomen, especially if marching is a necessity.

WOOLEN CLOTHING.—The following very interesting and useful suggestions on the importance of woolen clothing we find in "Hall's Journal of Health:"

"The most healthful clothing for our climate, the year round, is that made of wool. If worn next to the skin by all classes, in summer as well as winter, an incalculable amount of coughs, colds, diarrhœas and fevers would be prevented, as also many sudden and premature deaths from croup, diphtheria, and lung diseases. Winter maladies can be prevented by the ability of a woolen garment to keep the natural heat about the body more perfectly, instead of conveying it away as fast as generated, as linen and flaxen garments do, as also cotton and silk, although these are less cooling than Irish linen, as any man can prove by noticing the different degrees of coolness on the application of a surface of six inches square of flannel, cotton, and linen to the skin, the moment the clothing is removed. The reason is, that wool is a bad conductor of heat, and linen is a good conductor.

"It is more healthful to wear a woolen garment next the skin in summer, because it absorbs the moisture of perspiration so rapidly as to keep the skin measurably dry all the time. It is curious to notice that the water is conveyed by a woolen garment from the surface of the body to the outside of the garment, where the microscope shows it condensed in millions of pearly drops, while in the experience of the observant it is evident that if a linen shirt becomes damp by perspiration, it remains cold and clammy for a long time afterwards, and unless removed will certainly cause some bodily ailment.

"In the night-sweats of consumption, or of any debilitated condition of the system, a woolen night-dress is immeasurably more comfortable than cotton or linen, because it prevents that sepulchral dampness and chilliness of feeling which are otherwise inevitable.

"The British Government makes it imperative that every

sailor in the navy shall wear a woolen flannel shirt in the hottest climates. The shrinkage of woolen garments in washing, whereby they become hard, impervious, and board-like, has prevented their more general use; but there are three ways of preventing this, to a greater or less extent: either let about one-fourth of the material be made of cotton, have it dyed red or some other color before it is woven, or if it is greatly preferred that it shall be white, exercise proper care in the process of washing. To prevent white woolen stockings from shrinking, have wooden stretchers made of the size and general shape of the foot, and let the stockings remain on them until perfectly dried; or, before rinsing the stocking, double it so as to fold at the heel, and lay the foot on the leg, then roll it tight and twist it crosswise.

"In washing all woolen garments, put them in very hot soap-suds, so as to be covered; then, when cool enough to allow the hands to be put in, simply press them about with the fingers or hands, and before taking the garments out, make the water for rinsing several degrees hotter than that from which they are to be taken; but instead of wringing the water out, or twisting them about in the water, raise the garments out of the water, up and down, a good many times, and then lay them over a line and let them drip dry. This process will to a considerable extent prevent fulling or shrinkage, and is worthy of being communicated to every person who expects to be a housekeeper."

CHECKING PERSPIRATION.—The following is also extracted from that excellent work, "Hall's Journal of Health:"

"Edward Everett, the finished scholar, the accomplished diplomatist, the orator, the statesman, the patriot, became overheated in testifying in a court-room, on Monday morning, went to Fanueil Hall, which was cold, sat in a draught of air until his turn came to speak,—'But,' said he, 'my hands and feet were ice, my lungs on fire. In this condition

I had to go and spend three hours in the court-room!' He died in less than a week, from this checking of perspiration. It was enough to kill any man.

"Professor Mitchell, the gallant soldier and the most eloquent astronomical lecturer that has ever lived, while in a state of perspiration in yellow fever, the certain sign of recovery, left his bed, went into another room, became chilled in a moment, and died the same night.

"If, while perspiring, or while something warmer than usual, from exercise or a heated room, there is a sudden exposure in stillness to a still, cold air, or to a raw, damp atmosphere, or to a draught, whether at an open window or door, or a street corner, an inevitable result is a violent and instantaneous closing of the pores of the skin, by which waste and impure matters, which were making their way out of the system, are compelled to seek an exit through some other channel, and break through some weaker part, not the natural one, and harm to that part is the result. The idea is presented by saying that the cold is settled in that part. To illustrate: A lady was about getting into a small boat, to cross the Delaware; but wishing first to get an orange at a fruit stand, she ran up the bank of the river, and on her return to the boat found herself much heated, for it was summer; but as there was a little wind on the water, the clothing soon felt cold to her. The next morning she had a severe cold, which settled on her lungs, and within the year she died of consumption.

"A stout, strong man was working in a garden in May. Feeling a little tired about noon, he sat down in the shade of the house, fell asleep, and woke up chilly. Inflammation of the lungs followed, ending, after two years of great suffering, in consumption. On opening his chest there was such an extensive decay that the yellow matter was scooped out by the cupful!

"Multitudes of women lose health and life every year in one or two ways: by busying themselves in a warm kitchen until weary, and then throwing themselves on a bed or sofa without covering, perhaps without a fire; or by removing the outer clothing, and perhaps changing the dress for a common one, as soon as they enter the house after a walk or a shopping. The rule should be invariably to go at once to a warm room, and keep on all the clothing at least five or ten minutes, until the forehead is perfectly dry. In all weathers, if you have to walk or ride on any occasion, do the riding first."

VACCINATION.—A late number of Doctor Hall's "Journal of Health," good authority, contains the following instructive article upon a subject which interests every one in this locality at the present time. We hope our readers will not fail to profit by the suggestion:

"The matter of small-pox impregnates the air immediately around the person or bedding of the patient, and any unvaccinated person who comes within ten or twelve feet of such person or the bedding is very sure to be attacked by the small-pox, and to have the pimples appear within a fortnight.

"In some cases vaccination wears out, and ceases to be a protection against the small-pox, and exposure to it gives varioloid. The longer it remains from small-pox after vaccination, the more severe the attack, if taken at all.

"Those vaccinated in infancy are most likely to have the varioloid between the ages of sixteen and twenty-five. This being so, a most important practical inference is to be drawn, that the occurrence of puberty diminishes the power of vaccination against infection; hence it becomes the imperative duty of every parent to have the child vaccinated on entering the fifteenth year. If it does not take, no harm has been done, and the chance of taking an odious and fearful disease has been removed. This re-vaccination should be repeated at twenty-five, especially if that at fifteen did not take.

"In order to fix upon the reader's mind a strong and clear idea of the value and necessity of a re-vaccination, a single fact will be stated. The Prussian Government, more than any other, enforces vaccination. In 1837, of forty-seven thousand soldiers re-vaccinated, the full effect took place in twenty-one thousand; and of the last, although the small-pox prevailed all over Prussia that year, not one single soldier took it.

"Re-vaccination should be intrusted to the family physician, who should be sacredly enjoined to procure the matter from the arm of one whom he knows to be the healthy child of healthy parents, so as to avoid, as far as possible, the introduction of hateful diseases into the constitution of the re-vaccinated. Every parent should place this article where it may be frequently seen."

SUDDEN DEATHS.—Doctor Hall, in his well-known "Journal of Health," says that very few sudden deaths which are said to arise from disease of the heart, do really arise from that cause. To ascertain the real origin of sudden deaths, the doctor says the experiment has been tried in Europe, and reported to a scientific congress held at Strasburg. Sixty-six cases of sudden deaths were made the subjects of a thorough *post-mortem* examination. In these cases only two were found who died from disease of the heart. Nine out of the sixty-six died from apoplexy, while there were forty-six cases of congestion of the lungs; that is, the lungs were so full of blood they could not work, there not being room enough for a sufficient quantity of air to enter to support life. The doctor goes on to enumerate the causes that may produce congestion of the lungs. They are—cold feet, tight shoes, tight clothing, costiveness, sitting still until chilled through, after being warmed by labor or a rapid walk, going too suddenly from a close and heated room into the cold air, especially after speaking, and sudden, depressing news operating on the blood. These causes of sudden death being known, avoidance of them may serve to lengthen our valuable lives, which would otherwise be lost under the verdict of a heart complaint. That disease is supposed to be incurable, and hence many may not take the pains they would to avoid sudden death, if they knew it lay in their power.

DIET.—The following is a list of articles of diet, with the time required for their digestion:

ARTICLES.	HOW COOKED.	TIME. h. m.
Rice	Boiled	1.00
Sago	"	1.45
Tapioca	"	2.00
Barley	"	2.00
Milk	"	2.00
*Milk	Raw	2.15
*Tripe	Boiled	1.00
Venison Steak	Broiled	1.35
Turkey	Roasted or boiled	2.30
*Goose	Roasted	2.30
*Pig, sucking	"	2.30
Lamb	"	2.30
Chicken	"	2.45
Eggs	Soft	3.00
*Eggs	Hard-boiled	3.30
*Eggs	Fried	3.30
*Custard	Baked	2.45
*Salmon	Boiled	1.30
Oysters	Raw	2.55
*Oysters	Stewed	2.30
Beef	Roasted	3.30
Beef Steak	Broiled	3.00
*Pork Steak	"	3.15
*Pork Steak, lean and fat	Roasted	5.15
*Pork, recently salted	Boiled	4.30
Mutton	Roasted	3.15
Mutton	Broiled or boiled	3.00
*Veal	Broiled	4.00
*Veal Cutlets	Fried	4.30
Fowls	Boiled	4.00
*Ducks	Roasted	4.00
*Butter	Melted	3.30

MEDICAL DEPARTMENT. 57

ARTICLES.	HOW COOKED.	TIME. h. m.
*Cheese, old and strong	Raw	3.30
*Soup, Beef, vegetables and bread	Boiled	4.00
*Soup, Bean	"	3.00
Soup, Barley	"	1.30
Soup, Mutton	"	3.30
Soup, Chicken	"	3.00
*Hashed Meat and Vegetables	Warmed	2.30
*Sausages, fresh	Boiled	3.20
*Heart, Animal	Roasted	4.00
*Beans	Boiled	2.30
Bread	Baked	3.30
Dumpling, Apple	Boiled	3.00
Apples	Raw	2.50
*Parsneps	Boiled	2.30
*Carrots	"	3.15
*Turnips	"	3.30
*Cabbage	"	4.30
Potatoes	"	3.30

Those marked with a star (thus *) should be avoided or eaten very sparingly by the invalid.

In addition to the articles of diet enjoined in the preceding list, dyspeptics had better omit the following articles:

New bread,
Hot biscuits,
Hot rolls,
Spiced pies,
Rich pastry,
Cream,
Fried fritters,
Fried potatoes,
Pork,
Veal,
Rich soups,
Salt fish,
Salt meat,
Smoked beef,

Salads,
Lettuces,
Cucumbers,
Radishes,
Turnips,
Cabbage,
Nuts,
Hickory nuts,
Walnuts,
Cocoa nuts,
Almonds,
Filberts,
Pickles,
Spices,

Smoked ham,
Stuffing of meats,
Do. of poultry and game,
Fat bacon,
Boiled salmon,
Sausages,
Mackerel,
Cheese,
Eels,
Hard eggs,

Coffee,
Chocolate,
Green tea,
Wine,
Beer,
Spirits,
Tobacco,
Snuff,
Cigars.

WHAT FAT AND LEAN PEOPLE MAY EAT AND DRINK, AND WHAT THEY SHOULD AVOID.—The following list of articles of diet has been compiled by these eminent physiologists, the editors of the "American Phrenological Journal," with certain modifications from a late publication entitled, "The Hand-Book of Dining." Judgment must be used in applying them to individual cases, and it would be especially well for the lean folks to remember that the list recommended to them presupposes unimpaired digestive powers. Individuals taking it as a general rule should omit such articles of food as they find their stomachs incapable of digesting, or that in any way disagrees with them:

What Fat Folks may Eat and Drink.—Lean beef, veal, and lamb; poultry, game, and fish, except salmon; eggs, dry toast, greens, cabbage, turnips, spinach, lettuce, and salad plants generally; tea and coffee, without sugar or cream.

What Fat Folks should Avoid.—Fat or potted meats; bread, as far as practicable, except the dry toast; biscuits, rice, arrow-root, sage, tapioca, macaroni, and vermicelli; puddings, and pastry of all kinds; custard, cheese, butter, cream, milk, and sugar; potatoes, carrots, parsneps, and beets; all sweet fruits; cocoa, chocolate, beer, and liquor of all kinds.

What Lean Folks may Eat and Drink.—Fresh beef and mutton; poultry and game; fresh fish of all kinds; soups, broth, and beef tea; eggs, butter, cheese, cream and milk; sweet fruits, jellies, sugar, and honey; bread, biscuits (not hot, however), custard, rice, tapioca, and other farinaceous substances in puddings and otherwise; potatoes, beans, peas, beets, parsneps, carrots, cauliflowers, asparagus, and sea kale; cocoa, chocolate, tea, coffee, and milk.

What Lean Folks should Avoid.—Salt meats of all kinds; salted fish, pickles, lemons, salads, and vinegar; acid drinks, very sour fruits.

Drowning.—*Rules to be observed for the Recovery of Persons apparently Drowned.*— The body should be moved with all speed to the nearest house, unless too far, and after being stripped and dried, it should be placed in bed, between blankets, the head being slightly raised, whilst hot bricks and bottles are being got ready to place to the feet and abdomen. Artificial respiration should be commenced, by means of a bellows, until a physician arrives with a proper apparatus. Mustard poultices should be applied to the abdomen, friction with hartshorn to the chest, and an injection administered, composed of turpentine and gruel or hot water and spirits. Hartshorn to the nose is useful. When signs of animation begin to appear, a teaspoonful of brandy and water should be given, but do not force it down the throat. It is foolish to attempt to bleed.

The late Doctor Valentine Mott gives the following directions: "Immediately as the body is removed from the water, press the chest suddenly and forcibly downward and backward, and instantly discontinue the pressure. Repeat this violent interruption until a pair of bellows can be procured. When obtained, introduce the muzzle well upon the base of the tongue. Surround the mouth with a towel or handkerchief, and close it. Direct a bystander to press firmly upon the projecting part of the neck (Adam's apple), and use the bellows actively; then press upon the chest, to expel the air from the lungs, to imitate the natural breathing. Continue this at least one hour, or until signs of natural breathing come on. Wrap the body in blankets, place it near a fire, and do every thing to preserve the natural warmth, as well as to impart an artificial heat, if possible. Every thing, however, is secondary to inflating the lungs. Send for a medical gentleman immediately, and avoid all friction until respiration shall be in some degree restored."

SMALL-POX. — *Cure for Small-Pox.* — If there is any truth in the following item, it deserves the serious consideration of medical men every where:

"The 'German Reformed Messenger' has received a letter from a friend in China, which states that a great discovery is reported to have been recently made by a surgeon of the English army in China, in the way of an effectual cure for small-pox. The mode of treatment is as follows: When the preceding fever is at its height, and just before the eruption appears, the chest is rubbed with croton oil and tartaric ointment. This causes the whole of the eruption to appear on that part of the body, to the relief of the rest. It also secures a full and complete eruption, and thus prevents the disease from attacking the internal organs. This is said to be now the established mode of treatment in China, by general orders, and is regarded as a perfect cure."

We append the following letters on the treatment of this loathsome disease:

"*Editor New York Herald,*— As there are still cases of small-pox among us, I have thought it might be the means of preventing those who have been exposed from taking it, to give an account of its treatment in my father's family. We moved from Connecticut to New York city, and the youngest brother took it while at play in the street. None of us knew it was the small-pox until he was broken out — down sick. We sent for a doctor, who said at once, 'It is the small-pox!' We were very much frightened—coming, as we did, from the country, and knowing nothing of it—for we had all been exposed. He said he cared no more for the small-pox than he did for a cold. He told us—'Get some black cohosh or snake-root, and some white-root, about equal parts, bruise it, make a tea of it, and drink about two tablespoonsful at a time, frequently throughout the day, and none of you will take it. Keep the boy who has it on low diet; don't let him have any thing greasy to eat, and keep him in a dark room, and he will get along well enough.' We did as directed. None of us took it, and the brother got well without any difficulty, notwithstanding he had it so bad that he was a frightful sight to look upon.   I. W. B."

To Prevent the Defacing Marks of Small-Pox.—The following interesting letter is addressed to the editor of the "New Orleans True Delta:"

"Whilst it is known to medical men that a total suppression of the eruption of small-pox will endanger the life of the sufferer, it is proved by experience that such a suppression can be undertaken on small portions of the body, for instance on the face, without harm.

"The pommade of Bendeloque, made of six parts of pitch (pix nigra), ten parts of yellow wax, and twenty-four parts of mercurial ointment will answer that purpose. When used it should be warmed, and all parts of the skin of the face covered, and kept covered continually for the space of four days. The salve, if it proves efficacious, must be applied on the first or second day the eruption has broken out. Four days after the application, when removed, the eruption will have remained papulous, and the spots will disappear in time, without leaving any marks behind.

"Having experienced the good effects from the administration of this remedy, we hope the press of this city will promulgate it by publication, and earn their thanks from the smiling faces of those whom it will have spared a frightful disfiguration. Respectfully yours,

"M. SCHUPPERT, M. D."

Small-Pox Remedy. — The small-pox remedy, which cured three thousand cases in England, taken in all stages of the disease, is so simple that it cannot be too widely disseminated. It is:

 Cream of tartar...................... ¾ ounce.
 Rhubarb ....13 grains.
 Cold water........................ 1 pint.

The dose is from a wine glass to half a pint. In severe cases the latter should be administered. In cases characterized by delirium, great benefit has been obtained by applying a bottle of hot water to the feet. Plenty of fresh air is important, and an out-door airing at the earliest period practicable is recommended. When applied in the earliest stage of the eruption, it is arrested, and suppuration prevented without any injurious result. The mixture should be well stirred or shaken immediately before giving it.

Another Remedy. — The "Palmer Journal," speaking of this loathsome disease and a cure for it, says: "About sixty cases of small-pox have been treated at the State Almshouse during the past three months, with a single fatal result, and that was in the case of a man who was taken there in the last stages of the disease, from a neighboring town. The remedy used in all cases was a tea made from a plant known in medicine as *carracenia purpurea* (familiarly called 'ladies' saddle of water cup'), the root of which is the remedial part. This remedy is a new discovery in medical science, and has been used with excellent effect elsewhere. It allays the fever and irritation caused by the formation of pustules — the latter dying away rapidly, leaving slight, if any, traces of the disease."

Another Cure.—The following prescription is vouched for by the "Eastport (Maine) Sentinel," as a cure for this disease:

"Give to the patient two tablespoonsful of a mixture of hop yeast and water, sweetened with molasses, so as to be palatable, equal parts of each, three times a day. Diet—boiled rice and milk, and toasted bread moistened with water and without butter. Eat no meat. Give catnip tea as often as the patient is thirsty, and physic when necessary. If the above treatment is strictly followed, no marks of the small-pox will remain."

Scarlet Fever and Small-Pox.—Doctor William Fields, of Wilmington, Delaware, gives publicity to the following recipe, which if faithfully carried out, will cure, he says, forty-five cases out of fifty, without calling on a physician:

"For scarlet fever, give to adults one tablespoonful of good brewers' yeast in three tablespoonsful of sweetened water, three times a day; and if the throat is much swollen, gargle with yeast, and apply yeast to the throat as a poultice, mixed with Indian meal. Use plenty of catnip tea, to keep the eruption out on the skin for several days.

"For small-pox, use the above doses of yeast three times a day, and a milk diet throughout the entire disease. Nearly every case can be cured without leaving a mark."

Doctor Richardson, an English chemist, says that iodine, placed in a small box, with a perforated lid, destroys organic poison in rooms. During the continuance of an epidemic small-pox in London he saw the method used with benefit.

DIPHTHERIA.—The "New York Express" publishes the following remedy for this dangerous affection, remarking that it has been adopted by some of the most eminent physicians, and has never yet been known to fail, when applied as soon as the first symptoms appeared:

"Diphtheria, in its early stages, may be recognized by any person of ordinary capacity, by two marked symptoms: the sensation of a boney or hard substance in the throat, rendering it difficult and painful to swallow; and a marked fœtor or unpleasant smell of the breath, the result of its putrefactive tendency. On the appearance of these symptoms, if the patient is old enough to do so, give a piece of gum camphor, of the size of a marrowfat pea, and let it be retained in the mouth, swallowing slowly the saliva charged with it until it is gone. In an hour or so give another, and at the end of another hour a third; a fourth will not usually be required, but if the pain and unpleasant breath are not relieved, it may be used two or three times more, at a little longer intervals, say two hours. If the patient is young, powder the camphor, which can easily be done by adding a drop or two of spirits or alcohol to it, and mix it with an equal quantity of powdered loaf sugar—or better, powdered rock candy—and blow it through a quill or tube into its throat, depressing the tongue with the handle of a spoon. Two or three applications will relieve. Some recommend powdered aloes or pellitory with the camphor; but observation and experience have satisfied us that the camphor alone is sufficient. It acts probably by its virtue as a diffusible stimulant and antiseptic qualities."

Another Cure.—A Pennsylvania correspondent writes that when diphtheria was so prevalent in portions of that State, he has known the following remedy to be most efficacious, and requests its publication :

" Make two bags that will reach from ear to ear, and fill them with ashes and salt; dip them in hot water, and wring them out so that they will not drip; then apply them to the throat; cover the whole with a flannel cloth, and change them as often as they become cool, until the throat becomes irritated, near blistering. For children it is necessary to put flannel cloths between the ashes and throat, to prevent blistering. When the ashes have been on a sufficient time, take a wet flannel cloth and rub it with Castile soap until it is covered with a thick lather; dip it in hot water, apply it to the throat, and change when cool; at the same time use a gargle made of one teaspoonful of Cayenne pepper, one of salt and one of molasses, in a teacupful of hot water, and when cool add one-fourth as much cider vinegar, and gargle every fifteen minutes until the patient requires sleep. A gargle made of Castile soap is good to be used part of the time."

A correspondent in Maine, sending the above remedy, says there had been a number of deaths from diphtheria till this remedy was used, since which all have recovered

Another Cure. — "We have received a recipe for the cure of diphtheria," says the "New York Tribune," "from a physician who says that of one thousand cases in which it has been used, not a single patient has been lost. The treatment consists in thoroughly swabbing the back of the mouth and throat with a wash made thus : Table salt, two drachms; black pepper, golden seal, nitrate of potash, and alum, each one drachm. Mix and pulverize; put into a tea-cup, which half fill with boiling water; stir well, and then fill up with good vinegar. Use every half hour, one, two and four hours, as the recovery progresses. The patient may swallow a little each time. Apply one ounce each of spirits of turpentine, sweet oil, and aqua ammonia, mixed, every four hours to the whole throat and to the breast bone, keeping flannel to the parts."

A Specific Remedy. — The following is furnished to the "Germantown Telegraph," by a lady of New England, who says it has cured thousands, and is infallible, if properly used. She says: "As soon as the patient discovers that the dreaded membrane is formed in the throat, let a live coal of either wood or anthracite be brought, on which drop tar, and while the smoke arises place over it the bowl of a common clay pipe, and inhale the same, allowing the smoke to pass through the mouth and out of the nostrils. Let this be done every hour until the membrane is utterly destroyed, which under this treatment has never failed to be the case. In connection with this, let the physician prescribe the requisite dose of chlorate of potash, to be dissolved in a tumbler of water, a teaspoonful of which must be frequently taken. After the disease is relieved — cured, rather — build up the patient's strength with generous viands, wines, etc."

A Simple Remedy. — A distinguished physician, who has had many cases of this malady to treat, gives in a published letter the following as the result of his experience:

"In the early stages of the complaint, which is always accompanied by a soreness and swelling of the throat, let the patient use a simple solution of salt and water, as a gargle, every fifteen minutes. At the same time moisten a piece of flannel with a solution of the same kind, made as hot as the patient can bear it, and bind it around his throat, renewing it as often as the gargle is administered, and in the mean while sprinkle fine salt between the flannel and the neck. Use inwardly some tonic or stimulant, either separately, or if the prostration be great, use both together. The treatment, as may be seen, is extremely simple, and if used in the earlier stage of the disease, will effect a complete cure."

Another Remedy. — Diphtheria is a very troublesome and dangerous disease. A very easy remedy has been found — one that will effect a speedy relief. Take a common tobacco pipe, place a live coal in the bowl, drop a little tar upon the coal, then draw the smoke into the mouth, and discharge it through the nostrils.

Great relief in diphtheria is also said to be found by putting Cayenne pepper into the sharpest vinegar, and dropping in live coals and inhaling the steam from a tea-pot. It gives strength to throw up the detached membrane, besides affording great relief in breathing.

A Sure Cure for this Terrible Disorder.—The following was published in the "Missouri Democrat," in introducing which the writer says: "It has been used by myself, and others to whom I have given it, in over one thousand cases, without a failure. It will *always cure*, if the treatment is commenced before the diphtheria membrane extends into the air tubes, which is known by the great difficulty of breathing, and restlessness. In such cases no remedy yet discovered will always effect a cure; but if the patient is watched, and this treatment is used in time, there is no danger. I sent this receipt to a friend of mine in Missouri, and he used it on himself, his family and neighbors with such wonderful success that he requested me to send it to you for publication, as this horrible disease is prevailing extensively in parts of his State. It is this:

    Golden seal, pulverized...............1 drachm.
    Borax, pulverized....................1 drachm.
    Black pepper, pulverized.............1 drachm.
    Alum, pulverized....................1 drachm.
    Nitrate of potash...................1 drachm.
    Salt, pulverized....................2 drachms.

Put all into a common-sized tea-cup or vessel, which holds about four ounces, and half fill it with boiling water, stir it well, and then fill up with good vinegar; it is fit for use when it settles. Make a swab by getting a little stick about the size of a pipe stem, notch one end, and wrap a string of cotton cloth around it, letting the cloth project about half an inch beyond the end of the stick, so as not to jag the mouth and throat, and fasten with a thread. Swab the mouth and throat well every half hour, if the case is bad, and every hour if not bad; when the patient gets better, every two hours; and when still better, two or three times a day until well, which will be from two to seven days. Touch every affected spot—the uvula, tonsils and fauces, the whole of the back part of the mouth and top of the throat, and let the patient swallow a little of the wash each time you swab. Swabbing causes no pain, though the patient will gag, and sometimes vomit; but swab well, and a feeling of relief will follow every swabbing.

"Let every patient have a separate swab and wash, as the disease is undoubtedly infectious. Keep the swab pure, by pouring what you use each time into another vessel, and also wipe off any matter or slime that may be on the swab every time you take it from the mouth.

"Rub the following liniment on the throat every three or four hours, and keep a flannel cloth around the neck till well:

    Spirits of turpentine .... 1 ounce.
    Sweet or linseed oil .... 1 ounce.
    Aqua ammonia .... 1 ounce.

"Mix, and shake before using each time. Keep the bowels regular with castor oil, and keep the patient in the house, but ventilate well.

"The diphtheria wash and liniment will be found sufficient in all cases, if taken in time; and should you mistake any other sore throat for diphtheria, you will effect a cure almost invariably, as I use this for all common sore throats. I have never lost a case, and many have told me that no money would induce them, in these diphtheria times, to be without the wash and liniment, and when a soreness of the throat is felt, it is used, and a cure is always effected.

                              "W. A. SCOTT, M. D.

"PALMYRA, Warren County, Iowa."

DIARRHŒA.—"Numerous requests have been made," says the "Philadelphia Inquirer," "to republish the recipe for diarrhœa and cholera symptoms, which we gave in our paper several weeks ago, and which was used by the troops during the Mexican war with great success, we give it below:

    Tincture of Cayenne pepper .... 2 drachms.
    Laudanum .... 2 ounces.
    Spirits of Camphor .... 2 ounces.
    Essence of Peppermint .... 2 ounces.
    Hoffman's anodyne .... 2 ounces.
    Tincture of ginger .... 1 ounce.

"Mix all together. Dose—a teaspoonful in a little water, or half a teaspoonful repeated in an hour afterwards, in a teaspoonful of brandy. This preparation will check diarrhœa in ten minutes, and abate other premonitory symptoms of cholera immediately. In the latter disease it has been used with great success, to restore reaction by outward application."

Another Cure.—The following receipt is published as an excellent one for the benefit of those who are afflicted with dysentery or diarrhœa:

"Take a paper of arrow-root, mix it with cold water to the consistency of paste, grate into it one nutmeg, put it into a pan, pour boiling water on it, and let it simmer over a slow fire until it is cooked, stirring it all the time it is being cooked, then put into it four tablespoonsful of loaf or crushed sugar.

"I warrant this to cure within thirty-six hours, if in the mean time the patient eats nothing else, avoids strong drink, and uses as little water as possible.

"DAVID HOSFORD."

Another Cure.—A correspondent of the "St. Louis Republican" gives the following recipe as an almost certain cure for that scourge of our summers — the diarrhœa. It is very simple, and within the reach of all. It should at least have a fair trial:

"Take about two tablespoonsful of good wheat flour; brown it in a pan until it gets the color of parched coffee; be careful not to let it burn; season it with very little salt and pepper and a very small quantity of butter; pour in a pint and a half of water, and let it boil down to a pint.

"I have used it myself, and seen it tried in a great number of cases, and have never known it to fail, even in the most aggravated cases."

Another Cure.—The following is vouched for by the "New York Aurora:"

"Even after all other remedies have failed, a certain cure will be found for it in rice water. Boil the rice, make the water palatable with salt, and drink it copiously while warm. Simple as it is, we never knew it to fail."

Another Cure.—Eat three or four green strawberry leaves.

Another Cure.—Make a tea of green or ripe (ripe being the best) black haws, sweeten to taste, and drink often.

Another Cure.—Parch half a pint of rice perfectly brown, then boil it till it is perfectly cooked, and eat it slowly. It is said to check it in a few hours.

Another Cure.—Put enough of loaf sugar into good brandy to make it pretty sweet, then stir it with a nearly red hot iron rod, and take from one to two tablespoonsful two or three times a day.

Another Cure for the same, or Bowel Complaint in Children.—A tea made of "ragweed" is said to be good.

Cure for Diarrhœa or Dysentery.—The following was discovered by Doctor Perkins, of Salem, Massachusetts, many years ago:

"Saturate any quantity of the best vinegar with common salt; to one large tablespoonful of this solution add four times the quantity of boiling water; let the patient take of this preparation, as hot as it can be swallowed, one teaspoonful once every half minute until the whole is drank; this is for an adult. The quantity may be varied according to the age, size, and constitution of the patient. If necessary, repeat the dose in six or eight hours. Carefully avoid keeping this preparation in vessels partaking of the qualities of lead or copper. The success of the remedy depends much on preparing and giving the dose as above directed. Keep the preparation hot until all is taken."

Cure for same or Cholera Morbus.—Take equal proportions of "number six," spirits of turpentine, essence of pepperment, essence of cinnamon, and to an ounce of the mixture add a lump of gum myrrh the size of a common thimble. Dose—for an adult, a teaspoonful every half hour, or oftener if necessary, until the disease is checked.

Another Cure for Diarrhœa or Summer Complaint.—Blackberry cordial is an excellent remedy.

DYSENTERY.—*An Infallible Remedy.*—Doctor Page, of Washington, communicates the following to the "Republican" of that city:

"The following simple remedy, long known in family practice, was recently tried in the camp of the New York Twenty-second Regiment, where there were from eighty to one hundred cases of dysentery daily, and with rapid cures in every case:

"In a tea-cup half full of vinegar dissolve as much salt as it will take up, leaving a little excess of salt at the bottom of the cup. Pour boiling water upon the solution until the cup is two-thirds or three-quarters full. A scum will rise to the surface, which must be removed, and the solution allowed to cool. Dose—a tablespoonful three times a day, till relieved.

"The *rationale* of the operation of this simple medicine will readily occur to the pathologist, and in many hundred trials I have never known it to fail in dysentery and protracted diarrhœa."

Another Cure.—The following recipe for the cure of diarrhœa and dysentery is said to be the best in use:

 African Cayenne pepper..............32 grains.
 Camphor...........................32 grains.
 Best Turkey opium, powdered.........16 grains.
 Bourbon whisky..................... 4 ounces.

Dose—one teaspoonful every four hours.

Another Cure—Which a gentleman from Baltimore, Maryland, recommends as having never failed to give relief:

 Rhubarb ....16 grains.
 Salts of tartar....................32 grains.
 Prepared chalk....................48 grains.
 Oil of spearmint .... 4 drops.
 Laudanum.........................20 drops.
 Soft water........................ 2 ounces.

Put into a phial, and shake well before using. Dose—for

a child from one to four years old, one teaspoonful; for a grown person, one tablespoonful—to either, three or four times a day, each dose to be sweetened with loaf sugar, and kept in a cool place, to prevent it souring.

Another Cure.—The "Middletown Republican" copies the following, and certifies to its good effect, as proved by many experiments:

"Take Indian corn, roasted and ground in the manner of coffee, or coarse meal browned, and boil in a sufficient quantity of water to produce strong liquid like coffee, and drink a teacupful, warm, two or three times a day. One day's use, it is said, will ordinarily effect a cure."

Another Cure.—Take newly churned butter, before it is washed or salted; clarify it over the fire, and skim off all the milky particles; add one-fourth brandy, to preserve it, and sweeten with loaf sugar. Let the patient, if an adult, take two tablespoonsful twice a day. The above is given as a certain cure.

Another Cure.—Put an ounce of salts of tartar into a quart of cider, and drink that quantity daily.

Another Cure.—Take a tumbler of cold water and thicken it with wheat flour to about the consistency of thick cream, and drink it. This dose to be repeated several times a day, or as often as the patient is thirsty. If not cured the first day, continue on the second. It is said this never fails.

Another Cure.—First take a dose or two of castor oil—enough to physic well; then put one tablespoonful of good tea into a quart of new milk, and boil it down to a pint; then take a tablespoonful of the milk every two hours until a cure is effected. This is said to be good.

Another Cure.—Put half a drachm of nitrate of silver into half a pint of water; then put from one to two tablespoonsful of this mixture into half a pint of warm water, and use it two or three times a day for an injection, and after it has returned from the bowels (which will be almost immediately) give an injection of two tablespoonsful of warm starch and from one-half to one teaspoonful of laudanum. The starch should first be boiled, and about as thick as cream. If the first injection is very painful, use only from one-half to one tablespoonful of the mixture. This is known to be very good.

NEURALGIA.—The following receipt is said to be a certain cure for this painful disease: Take two large tablespoonsful of Cologne water, and two tablespoonsful of fine salt; mix them together in a small bottle. Every time you have any acute affection of the nerves, or neuralgia, simply inhale the fumes from the bottle through your nose, and you will be immediately relieved.

Another Cure.—Treating on this disease, the "Lawrenceville Herald" says: "As this dreadful disease is becoming more prevalent than formerly, and as the doctors have not discovered any method or medicine that will permanently cure it, we simply state that some time past a member of our family had suffered most intensely from it, and could find no relief from any remedy applied, until we saw an article which recommended the application of bruised horseradish to the face for toothache. As neuralgia and toothache are both nervous diseases, we thought the remedy for one would cure the other, so we made applications to the side of the body where the disease was seated, and it gave almost instantaneous relief to the severe attack of neuralgia. Since then we have applied it several times, and always with the same gratifying results. The remedy is simple, cheap, and within the reach of every one."

Another Cure.—Doctor Caminiti, of Messina, appears to have discovered a valuable remedy for certain neuralgic pains. A lady, a patient of his, had long been suffering from trifacial neuralgia. She could not bear to look at luminous objects; her eyes were continually watering, and she was in constant pain. Blisters, preparations of belladonna, hydrochlorate of morphine, friction with the tincture of aconite, pills of acetate of morphine and camphor, sub-carbonate of iron, etc., had been employed with but partial success, or none whatever. At length, Doctor Caminiti, attributing the obstinacy of the affection to the variation of temperature so frequent in Sicily, hit upon the plan of covering all the painful parts with a coating of collodion, containing hydrochlorate of morphine in the proportions of thirty grammes of the former to one of the latter. The attempt was perfectly successful, and the relief was instantaneous and permanent, and the coating fell off in the course of a day or two.

Another Cure.—Good whisky and quinine, it is said, will cure this disease.

ERYSIPELAS.—The editor of the " Salem Observer " gives a public cure for this disorder, from which he has been a great sufferer. He says: " A simple poultice made of cranberries pounded fine, and applied in a raw state, has proved in my case, and that of a number of persons in this vicinity, a certain remedy." In this case the poultice was applied on going to bed, and the next morning, to his surprise, he found the inflammation nearly gone, and in two days he was as well as ever.

Another Cure.—A doctor who has retired from practice sends the following to a San Francisco paper for publication, saying: " Please give your readers the benefit of this, for two persons have already died with erysipelas produced by cold in the wounded part: Take the common yellow carrot, scrape or grate it fine, and apply as a poultice. It is a sure cure. The same for croup in children. Apply to the neck and breast, changing the poultice when it becomes dryish."

Another Cure.—Make egg wine, rich and good for drinking; drink part of it, and wash the diseased place with the remainder. This is infallible.

Another Cure.—A poultice of poke-root, bruised, cooked, and applied, is said to be good.

Another Cure.—Rye flour, put on dry, and let remain on twenty-four hours; then wash off with warm Castile soap suds, and apply again as before, and so continue, and use a little flour of sulphur and nitre.

Another Cure.—Cover the swelled part with a cloth made very wet with a *strong* solution of sugar of lead, and change the cloth every five minutes until the swelling disappears, and physic well until a cure is effected.

DYSPEPSIA.—Drink no cold water within half an hour of eating; use light, dry diet, keep the skin clean, and brace up the system by copiously breathing pure fresh air.

Another Cure.—A teaspoonful of Epsom salts and the same quantity of magnesia, in a glass of cold water, every morning, on an empty stomach.

Another Cure.—Ground cinnamon one part, aloes one part, copperas one-fourth part, and opium one-sixteenth part—all mixed together. Dose—half a teaspoonful in sugar every morning and evening.

Another Cure.—A teaspoonful of white mustard seed, whole, taken two or three times a day, is pronounced to be an excellent remedy.

Another Cure, in Bitters.—Half fill a jug with wild cherries, and then fill it with old Jamaica spirits. Dose—half a wine glass twice a day, without sugar. It is very strengthening.

Another Cure, in Bitters.—

Centaury	5 ounces.
Senna	3 ounces.
Boletus of oak	4 ounces.
Canella alba	4 ounces.
Caraway seed	2 ounces.
Rhubarb	2 ounces.
Gum myrrh	2 ounces.

The last three to be pulverized, and all to be put into one gallon of rum, and then let stand a week. Dose—one tablespoonful fifteen minutes before each meal.

For Weakness and Gnawing of the Stomach, Dyspepsia, and General Debility.—

Socotrine aloes........................1 ounce.
Rhubarb, pulverized....................1 ounce.
Cassia senna...........................1 ounce.
Cloves, pulverized.....................1½ ounce.
Cinnamon, pulverized...................1½ ounce.
Allspice...............................2 ounces.

Add two nutmegs, pulverized. Put all into half a gallon of good whisky, and let it stand nine days; then draw off the liquor, and pour on the spices three pints of water; shake them up well, and let them stand a few hours; then draw off the water and mix it with the liquor, adding to it two pounds of white sugar.

Dose—from one to two teaspoonsful three times a day, immediately before eating.

Dyspeptic Ley.—The following was employed by the eminent Doctor Physic, in his own case, and, we are informed, was of decided advantage when all other remedies failed: Take of hickory ashes one quart, soot six ounces, boiling water one gallon; mix and stir frequently. At the end of twenty-four hours pour off the clear liquor. A tea-cup may be taken three times a day.

CONSUMPTION.—The following recipe was furnished me through the kindness of the Reverend Edward A. Wilson, who was cured of consumption by the use of this prescription. His address is Williamsburg, King's county, New York:

Recipe for Consumption, Asthma, Bronchitis, Scrofula, etc.—

Extract blodgetti......................3 ounces.
Hypophosphite of lime..................½ ounce.
Alatin (pura)..........................1 drachm.
Meconin (pura).........................½ scruple.
Extract cinchona.......................2 drachms.

Loaf sugar	1 pound.
Pure port wine	½ pint.
Warm water	1 quart.

To prepare the above recipe properly, all the powders and extracts should be thoroughly compounded and mixed well together, and placed in a vessel or bottle holding at least three pints; then pour into the bottle about half a pint of warm water, and shake it well, which will turn the whole a bright red color. Let it stand a few moments, then add the other pint and a half of warm water, with the sugar dissolved in it; also, add the wine (or, if you have no wine, rum or Holland gin will do). Shake well, and when cold it is ready for use. Dose—one large tablespoonful four time a day. Shake the bottle each time before use. Keep the bottle in a cool place, and in no case allow it to stand in a room with a fire.

Another Cure.—The following recipe was furnished me through the kindness of the Reverend William Cosgrove, of Brooklyn, New York, who was cured of consumption, while laboring as a missionary in Japan, by the use of this prescription, which was obtained from a learned physician in the city of Jeddo:

The Japanese Recipe, for the cure of Consumption, Bronchitis, Throat Complaints, Coughs and Colds, and the Debility of Constitution which these Disorders bring on.

Bark of the re-kaila arborea, in powder	4 drachms.
Bark of the boughi nepeuthes, in powder	3 drachms.
Extract of gauranin	½ drachm.
Extract of veronica gelatina	2 drachms.
Extract of pyritis	1½ drachm.
Sugar	¾ pound.

Mix these ingredients well together, and moisten them with four tablespoonsful of hot water; then pour on one pint of cold water; stir the mixture, and then pour it into a bottle, and add half a pint of sherry wine or brandy.

Cork the bottle tight, and keep it in a cool place. The medicine will be ready for use in two hours after it is mixed. The bottle should be shaken each time before taking a dose.

Dose—one tablespoonful every four hours, at first; if the pain and cough are very severe, the dose may be taken a little oftener. If the cough is troublesome, and the patient restless at night, half a tablespoonful may be given every two hours until the patient sleeps.

Mr. Cosgrove remarks: "In the pamphlet which I am about to publish I will give a full description of the different ingredients in this recipe. I will merely remark in this place that the re-kaila arborea is a perennial plant, growing abundantly in Japan and its islands, also in parts of China and Tartary. The bonghi nepenthes is a small tree which grows in Japan, Asia, and also Turkey in Europe. The extracts of gauranin, veronica, gelatina and pyritis, are vegetable products obtained from various plants. Though they have been long known to the Japanese physicians as invaluable remedial agents, it is but lately that European chemists have been able to prepare them properly. The Japanese physicians give this compound medicine the name of the Re-Kaila Mixture, from the first and most important of the medicines which compose it, namely, the re-kaila arborea."

An Indian cure for the same.—Take of the barks of black oak, white oak, beech, sycamore, sweet gum, poplar and wild cherry, and of the roots of dogwood, sassafras and sarsaparilla, of each a handful, and make a strong tea; then add elecampane, hoarhound, hyssop, sage and camomile, of each a handful, and boil down to one gallon, and strain it; then add one quart of honey, and one pint of vinegar, and a tablespoonful of anvil dust, and then simmer it half an hour. Dose—half a teacupful three times a day. Drink strong tea made of red chips of sycamore, diet lightly, abstain from hog meat and sweet milk, and be careful not to take cold.

Another Cure.—Take one pint of good spirits and one pint of pine tar, and mix them together in a stone jug or crock, and let it stand twenty-four hours; then add one pint of strong hoarhound tea and one pint of honey; then simmer it awhile over a slow fire and skim off the froth; then bottle up the liquor, and take a teaspoonful morning, noon and night.

Another Cure.—Take an equal quantity of the roots of elecampane, spikenard, sarsaparilla and burdock, and make a strong tea; and then make it into a syrup with honey, and drink a little of it three or four times a day. This is very good.

Another Cure.—Pine tar eaten on light wheat bread (same as butter) is said to have cured this disease when other remedies had failed. A tea made of St. John's wort is very good for constant drink in this disease. Try it. And also, a small quantity of the juice of green hoarhound, mixed with half a pint of new milk, and drank warm every morning, is said to be very good.

For a Consumptive Cough or Pain in the Breast.—Take one tablespoonful of pine tar, three tablespoonsful of honey, and three yolks of hen's eggs, and beat them well together; and then add half a pint of wine, and beat all well together in a dish with a spoon; and then bottle it up for use.

Dose—a teaspoonful morning, noon and night, before eating. Drink a tea made of hoarhound, St. John's wort, or barley, for constant use.

The Geography of Consumption.—The following valuable information is copied from the manuscript of a forthcoming work entitled " Influence of Climate in North America," to be compiled by Mr. J. Disturnell:

" Consumption originates in all latitudes from the equator where the mean temperature is eighty degrees Fahrenheit, with slight variations, to the higher position of the temperate zone, where the mean temperature is forty degrees, with sudden and violent changes. The opinion long entertained, that it is peculiar to cold and humid climates, is founded in error. Far from this being the case, the tables of mortality warrant the conclusion that consumption is sometimes more prevalent in tropical than in temperate countries. Consumption is rare in the Arctic regions, in Siberia, Iceland, the Orkneys and Hebrides; also in the north-western portion of the United States.

"In North America 'the diseases of the respiratory organs, of which consumption is the chief, have their maximum in New England, in latitude about forty-two degrees, and diminish in all directions from this point inland. The diminution is quite as rapid westward as southward, and a large district near the fortieth parallel is quite uniform at twelve to fifteen per cent. of deaths from consumption, while Massachusetts varies from twenty to twenty-five. At the border of the dry climate of the plains in Minnesota, a minimum is attained as low as that occurring in Florida, and not exceeding five per cent. of the entire mortality. It is still lower in Texas, and the absolute minimum for the continent in temperate latitudes is in Southern California.'

"The upper peninsula of Michigan, embracing the whole of the Lake Superior region, Minnesota, Nebraska and Washington Territory, are all alike exempt, in a remarkable degree, from the above fatal disease. Invalids suffering from pulmonary complaints and throat disease are almost uniformly benefited by the climate of the above northern region, having a mean annual temperature of from forty to forty-five degrees Fahrenheit."

Coughs and Colds.—A recipe for making the best cough syrup:—Take one ounce of thoroughwort, one ounce of slippery elm, one ounce of stick licorice, and one ounce of flaxseed. Simmer them together in one quart of water, until the strength is entirely extracted, then strain carefully, and add one pint of best molasses, and a half-pound of loaf sugar; simmer them all together, and when cold bottle up tight for use. This is the cheapest, best, and safest medicine for coughs in use. A few doses, of one tablespoonful at a time, will alleviate the most distressed lung cough. It soothes and allays irritation, and if continued in use, it will subdue any tendency to consumption. It breaks up entirely the hooping-cough, and no better remedy can be found for croup, asthma, bronchitis, and all affections of the lungs and throat.

Another Remedy.—

Black cherry bark .... 1 ounce.
Squills................................ 1 ounce.
Seneca snake-root..................... 1 ounce.
Blood-root .... 1½ ounce.

The above to be put into three pints of warm water; steep four hours, then strain. Add quarter of a pound of loaf sugar, then steep down to half a pint and bottle.

Dose — for adults, a large teaspoonful three or four times a day, to be increased or diminished as the case may require.

For a Hectic Cough.—Take three yolks of hens' eggs, three spoonsful of honey, and one spoonful of tar; beat them well together, and add to them one gill of wine. Take a teaspoonful three times a day, before eating.

Another Cure.—Take half a pound of wild licorice, half a pound of brook liverworth, two ounces of elecampane, four ounces of Solomon's seal, half a pound of spikenard, and four ounces of comfry, boiled in four quarts of water, to which add two pounds of honey and one pint of old spirits. Half a wine glass before eating is a dose.

Cure for a Cold. — Take a teaspoonful of fine salt in the mouth, and swallow it as dry as possible on going to bed, and then drink half a tumbler of cold water.

Cure for a Cough. — One tablespoonful of molasses, two teaspoonsful of castor oil, one teaspoonful of paragoric, and one of spirits of camphor. Mix, and take often. The editor of the "Farmer" says of this recipe: "It was prescribed for us when we were suffering from a cough that seemed as if we were on the brink of consumption—no cessation nor rest day or night. We took it, and were cured in three days."

Dry Cough.—Take of powdered gum arabic half an ounce; dissolve the gum first in warm water, squeeze in the juice of a lemon, then add two drachms of paregoric and one of syrup of squills. Cork all in a bottle and shake well. Take one teaspoonful when the cough is troublesome.

Thompson's Cough Syrup.—Take of poplar bark and bethroot, each one pound, and of water nine quarts; boil them gently in a covered vessel for fifteen or twenty minutes, then strain through a coarse cloth; add seven pounds of loaf sugar, and simmer till the scum ceases to rise. When the syrup is nearly cold add one pint of tincture of lobelia and one gallon of pure French brandy. Dose—a tablespoonful three or four times a day.

Cure for Inveterate Coughs.—Tea made of coltsfoot and flaxseed, sweetened with honey, is a cure for inveterate coughs. Consumptions have been prevented by it.

Coughs.—It is said that a small piece of rosin dipped in the water which is placed in a vessel on the stove (not an open fire-place) will add a peculiar property to the atmosphere of the room, which will give great relief to persons troubled with a cough. The heat of the stove is sufficient to throw off the aroma of the rosin, and gives the same relief that is afforded by the combustion, because the evaporation is more durable. The same rosin may be used for weeks. Flaxseed syrup is also an excellent remedy for a cough.

Another Cure.—Make a tea out of the leaves of the pine tree, and sweeten it with loaf sugar; drink it freely warm on going to bed, and cold through the day. Try it.

For Cough or Croup.—Slice onions and spread brown sugar over the slices; put them into a pan, set them in a stove to stew, and when they are soft squeeze out the juice. Dose—

from a few drops to a teaspoonful, according to the age, several times a day, if the cough is bad. For croup give a teaspoonful.

Another Cure.—Make a strong tea of hoarhound, liverwort, ground ivy and licorice root; sweeten it with honey and loaf sugar, to make a good syrup, and take a tablespoonful, if an adult, often, if the cough is troublesome.

HOOPING-COUGH.—A teaspoonful of castor oil to a tablespoonful of molasses; a tablespoonful of the mixture to be given whenever the cough is troublesome. It will afford relief at once, and in a few days will effect a cure. The same remedy relieves the croup, however violent the attack.

Another Cure.—The best kind of coffee, prepared as for the table, and given as a common drink to the child, as hot as it can be taken, and a piece of alum for the patient to lick as often as he may wish. Most children are fond of alum, and will get all they need without being urged; but if they dislike it, they must be made to taste it eight or ten times in the course of the day. It will effectually break up the worst case of hooping-cough in a very short time. To adults or children in the habit of taking coffee the remedy is good for nothing.

Another Cure.—Put five cents' worth of wild cherry bark and five cents' worth of licorice into a pint of water, and boil down to one-half, then put in five cents' worth of paregoric and one-eighth of a pint of good brandy. Dose.—Commence by giving small doses, according to the age, three times a day.

Another Cure.—Take twenty grains of salts of tartar, ten grains of cochineal, and one ounce of refined sugar; dissolve them in one gill of warm water. Dose—for an infant give a teaspoonful, morning, noon, and night, and a little every time the cough is troublesome.

CROUP.—" Hall's Journal of Health" contains the following advice: "This is an inflammation of the inner surface of the windpipe. Inflammation implies heat, and that heat must be subdued, or the patient will inevitably die. If prompt efforts are made to cool the parts in the case of an attack of the croup, relief will be as prompt as it is surprising and delightful. All know that cold water applied to a hot skin cools it, but all do not understand that hot water applied to an inflamed skin will certainly cool it off; hence the application of iced water with linen cloths, or almost hot water with woolen flannel, of two folds, large enough to cover the whole throat and upper part of the chest. Put these into a pail of water as hot as the hand can bear, and keep it thus hot by adding water from the boiling tea-kettle. Let two or three flannels be in hot water all the time; and one on the throat all the time, with a dry flannel covering the wet one, so as to keep the heat in to some extent. The flannels should not be so wet when put on as to dribble, for it is important to keep the clothing dry, and keep up the process until the phlegm is loose, and the child is easier, and begins to fall asleep; then gently wrap a flannel over the wet one which is on, so as entirely to cover it, and the child is saved. When it awakes both flannels will be dry."

Another Cure.—Doctor Goodman recommends the following as a simple yet certain remedy for this common and often fatal disease among children. He says: "Whenever they are threatened with an attack of croup, I direct a plaster covered with Scotch snuff, varying in size according to the age of the patient, to be applied directly across the thorax, and retained there till all the symptoms disappear. The remedy is always found to be effectual when applied in the first and second stages of the malady." The plaster is made by greasing a piece of linen and covering it with snuff.

Another Cure.—A piece of fresh lard, as large as a butternut, rubbed up with sugar, in the same way that butter and sugar are prepared for the dressing of puddings, divided into three parts, and given at intervals of twenty minutes, will relieve any case of croup not already allowed to reach the fatal point.

Another Cure.—As this is a very dangerous and rapid disease, the best medical aid should be procured as soon as practicable. In the mean time the most strenuous efforts should be made to arrest the progress of the disease. Bathe the feet in hot water, and put draughts on the feet, with mustard on them. Simmer onions with lard, and apply to the throat. A piece of linen or cotton cloth soaked in lard or sweet oil, sprinkled over with Scotch snuff, and applied where the disease is greatest, will often afford relief. Turpentine mixed with hot water, a flannel cloth dipped into it and applied to the throat, and the hands and feet rubbed with it, is a good remedy for this distressing complaint. Hive syrup taken internally, or a syrup made of sliced onions and white sugar, will often be found effectual in arresting the progress of the disease.

Another Cure.—Croup should have immediate relief when first discovered, or it will soon become incurable, and prove inevitably fatal. It is caused by the formation of a false membrane across the windpipe, which must either be prevented, absorbed, or ejected, else suffocation is certain to ensue. As soon as the peculiar rattle in the throat is heard, an emetic should be given (antimony, if convenient), and when that has operated, frequent doses of cherry pectoral will subdue the disease.

Another Cure.—Take out of a large-sized onion about as much of the heart as a hazelnut, and fill the cavity with flour of sulphur; then wrap up the onion in brown paper, and roast it until soft; then squeeze out the juice, and add to it three teaspoonsful of molasses and one teaspoonful of pulverized alum. Dose—a teaspoonful every twenty minutes until the child vomits. This is said to be very good.

Another Cure for Croup or Hives.—Roast a large onion well, squeeze out the juice and sweeten it with honey until it becomes a thick syrup, then add two drops of spirits of turpentine. This can be given to a child from six months to a year old, through the course of the day. Do not allow the child to go out in the wet or damp air. This is very good; try it.

CANCER.—Mr. Thomas Tyrell, of Missouri, advertises that a cancer on his nose, which had been treated without success by Doctor Smith, of New Haven, and the ablest surgeon in the western country, had been cured in the following manner: He was recommended to use strong potash made of ashes of red oak bark, boiled down to the consistency of molasses, to cover the cancer with it, and in about an hour afterward to cover with a plaster of tar, which must be removed after a few days, and if protuberances remain in the wound, apply more potash to them and the plaster again, until they disappear, after which heal the wound with common salve. Cautery and the knife had been previously used in vain. This treatment effected a perfect and speedy cure.

Another Cure. — A writer in the "Philadelphia Evening Journal" claims to have an infallible cure for cancer. The recipe is: The juice of the sheep-sorrel pressed and exposed on a pewter plate in the sun, until somewhat jellied. Apply it on the skin over and around the cancer, the application to be continued until the cancer and its roots loosen and drop out, which will be in the course of three or four days. The ingredients of which the pewter is composed combining with the acid of the plant, are believed to be important in the compound. The leaves of the sheep-sorrel are what botanists call *sagittate*, which is resembling in shape the head of an arrow. The writer also states that he cured his corns by an application of the leaves of the sheep-sorrel to them, which in a few hours softened them so much that they peeled off, and a cure was effected.

Another Cure.—It has been ascertained that the application of raw cranberries, applied as a poultice, will cure this most inveterate disease. "We know of one instance, a lady of our acquaintance," says an exchange paper, "who had a cancer in her breast, which had become as large as a pullet's egg, and was an inch below the surface of the skin. In this present case it was an hereditary disease, and she regarded it as a death-warrant. She was persuaded, however, to try the cranberries, and they effected a cure. It is now between two and three years since it disappeared, and she has had no intimation of a return of the disease. The cranberries were mashed in a mortar, spread on a cloth, and laid on, changing the poultice three times a day. In two or three days it became so sore it drew out pustules, that filled like the small-pox, and this process was renewed with the same effect until the whole was drawn away, the cancer becoming softened and decreasing in size at every application, until it finally disappeared."

Another Cure. — Put a sticking-plaster all around the cancer, leaving a hole a little larger than the cancer; then make a plaster of chloride of zinc, blood-root and wheat flour, and spread it on a piece of linen or muslin of the size of this opening, and apply it to the cancer. Let it remain on for twenty-four hours; then remove it, and if the cancer cannot then be taken out, put on another plaster of the same, and let it remain for twenty-four hours, or until the cancer can be removed. Continue this course until the cancer is killed, then dress the wound with some healing salve.

CANCER WART.—Make a salve of the yolk of a hen's egg and as much alum salt, and after having shaved off the outer skin, apply the salve, and let it stay on until it comes off itself; then dress it with a salve made of poke leaves, by drying them in the sun, on a pewter dish.

FELON.—The following, which is taken from the "Buffalo Advertiser," is said by some one, who pretends to know all about it, to be a sure remedy for a felon. Take a pint of common soft soap, and stir it in air-slaked lime, till it is of the consistency of glaziers' putty. Make a leather thimble, fill it with this composition, insert the finger therein, change the composition once in twenty minutes, and a cure is certain. We happen to know that the above is a certain remedy, and recommend it to any who may be troubled with this disagreeable ailment.

Another Cure.—Dissolve in boiling hot water in a tea-cup, a tablespoonful of impure carbonate potassa; when cold wet a cloth and apply it to the part affected; let it be kept wet with the solution till pain and soreness is gone, which will be sooner or later, as regards the progress the disease has made when applied. A pure article of salæratus is a good substitute, if impure carbonate potassa cannot be obtained. This is a sure cure.

Another Cure.—When a felon appears on the hand, apply a piece of rennet soaked in milk to the affected part, and renew the application at brief intervals until relief is found. The remedy may be obtained of any butcher. This article was first recommended by a skillful physician, now deceased. It has been tried in many cases, and has never failed to afford relief.

Another Cure.—As soon as the part begins to swell, get the tincture of lobelia, and wrap the part affected with a thick cloth saturated thoroughly with the tincture, and the felon is dead. An old physician says he has known this to cure in scores of cases, and that it never fails if applied in season.

Another Cure.—Blue flag-root and wild turnip-root, a handful of each stewed in half a pint of hog's lard; strain it off, add to it four spoonsful of tar, simmer them together; apply this ointment to the felon till it breaks. Add beeswax and rosin to the ointment for a salve to dress it with after it is broken. This is an infallible cure without losing a joint.

For Felon or Run-Round.—If a felon or run-round appears to be coming on the finger, you can do nothing better than to soak the finger thoroughly in hot ley. It will be painful, but it will cure a disorder much more painful.

Another Cure.—Take an equal quantity of red lead and Castile soap, and make them into a salve with ley, and apply it.

SNAKE BITE.—The "Cultivator," published at Indiana, says that alum is a sure antidote to the bite of a rattlesnake. Take a piece about the size of a walnut, and dissolve and drink it. This it is said, will cure either man or beast. It would be a good idea for those who are in the habit of going into the woods or the prairies, to carry a piece of alum in their pockets, so as to be in readiness for the emergency of a bite. It is a simple remedy, but we have no doubt of its efficacy.

Another Cure.—The "New Albany Bulletin" says: "We learn that a lady resident of this county was bitten by a snake of the copper-head kind, in the foot, a few days since, which caused such excruciating pain as to give her friends serious apprehension, for a time, of her recovery. An experiment was, however, resorted to, which resulted in a cure. The snake was killed, cut to pieces, and bound to the wound, which in a short time extracted the poison, and the lady is now convalescent."

Cure for Rattlesnake Bites and other Poisonous Creatures.—Indigo four drachms, gum camphor eight drachms, alcohol eight ounces; mixed and kept in close bottles. Apply to the wound and the cure is soon completed.

Another Cure.—Take green hoarhound tops, pound them fine, press out the juice, let the patient drink a tablespoonful of the juice morning, noon and night, or three times in twenty-four hours. Apply the pounded herbs to the bite, change the same twice a day. The patient may drink a spoonful of sweet olive oil. This never fails to cure.

Another Cure.—If the bite is on a limb, instantly tie a cord tightly above the part bitten, and then apply a cupping-glass on the bite, and bathe it with spirits of hartshorn. Take a dose of sweet oil, drink whisky freely, and take a tablespoonful of the juice of the tops of green hoarhound three times a day. Certain cure.

PUTRID SORE THROAT.—A lady who has experienced the benefit of the following simple remedy is very anxious that others should be made acquainted with it and its value:

Mix one gill of strong apple vinegar, one tablespoonful of drained honey, and half a pod of red pepper, or half a teaspoonful of ground pepper; boil them together to a proper consistency, then pour it into half a pint of strong sage tea. In severe cases, half a spoonful for an adult. As the canker decreases, decrease the frequency of the dose.

Another Cure.—Mix a penny's worth of pounded camphor with a wine glass of brandy, pour a small quantity upon a lump of sugar every hour, and allow it to dissolve in the mouth. The third or fourth time enables the patient to swallow with ease. This has cured in the last stages.

Chlorate of Potash.—Some one says: "Every body should keep a quantity of chlorate of potash. We have never found anything equal to it for a simple ulcerated sore throat. Dissolve a small teaspoonful of it in a tumbler of water, and occasionally use a spoonful of the solution as a gargle for the throat. It is nearly tasteless, and not at all offensive to take, hence it is well adapted to children. Nothing is better than this for chapped or cracked hands. Wash them in a weak solution, and they will soon be well. It is also good for a rough, pimply, or chapped face.

Sore Mouth.—The best local remedy for sore or ulcerated mouth or throat is the frequent application of the tincture of nutgalls, diluted with an equal portion of cold water, or a tea made of the galls may be substituted, and when cold wash or rinse the mouth with it very frequently.

Wash and Gargle for Sore Mouth and Throat.—Take of the blackberry root and gold thread each one ounce, sage two ounces, rose leaves half an ounce, water two pints; boil down to one-half, and strain; add one pint of honey, and boil down to one pint; add, while hot, alum and borax, of each a piece the size of a cranberry. This is known to be a sure remedy for nursing sore mouth or thrush.

Another Cure. — Take a lump of nitre (saltpetre) the size of a common thimble, twice that of alum, and half a gill each of charcoal and yellow-root (yellow pacoon — golden seal), all pulverized and mixed together, and made very wet with honey. Eat of it, and rub it on the part affected, and if the throat is too sore to swallow any of it, pour on some water, mix it well, let it stand awhile, and then strain out the water and gargle with it. This is known to be an excellent remedy for this disease.

Another Cure.— Take two tablespoonsful of red pepper and one tablespoonful of table salt, and pour on a pint of water and vinegar, in equal parts, boiling hot, and let it stand an hour. Dose—a tablespoonful every half hour, and gargle with the same. This, too, is an excellent remedy for this disease, and also for quinsy.

WHITE SWELLING.—Having suffered exceedingly in my youth from what was called (though perhaps improperly so) a white swelling, I will give a brief statement of what I believed to have been the cause thereof, and which is probably the cause of like suffering in others, and also the mode of treatment.

I had hopped for a great distance, both up and down hill, with quite a load upon my shoulders, and in one hour thereafter I felt a pain in the knee upon which I had done the hopping. The pain first appeared to be in the joint, but probably it was not there, but just above it, and may have been caused by bruising the joint, or over-straining some ligament, cartilage, muscle or membrane near it, in hopping.

Hopping was the primary cause, but whether that produced strain or bruise as the immediate cause of the pain, I am unable to tell; but I believe one or both of these things were the immediate cause of my suffering, and are usually the cause of what are commonly called white swellings, though perhaps they are the effect of a scrofulous taint in the system. They may also be produced by severe local colds, and may sometimes be seated in the marrow. But whatever the cause, they require the most prompt and thorough treatment. And it is the imperative duty of parents and guardians to use every means in their power to arrest the disease in the beginning, for by doing so they may thereby save the sufferer from untold misery and lameness through life.

I know not what would be the proper treatment. That would, perhaps, depend to some extent on the cause that produced the complaint. As a general thing I would advise bleeding, physicking, dieting, and *resting;* and that the part affected be thoroughly sweated or fomented with chamomile flowers, hops, or the like, and kept very warm with flannels, and with poultices of the most scattering kind. I know of nothing better for such a poultice than the roots of a weed or plant called square-stalk or carpenters' square. Hops and mullen leaves are also good.

If you sweat and poultice, be extremely careful for a long time not to take cold, for if you should, the pain will most likely return and be ten-fold worse than before — rending the bone in splinters like lightning does the oak.

Should the pain not yield to sweats and poultices within a short time, I would bathe the part with the most scattering or driving liniments or embrocations that I could get. I know not the best, but if I could get nothing that I thought better, I would try Davis's pain-killer, Mexican mustang, or nerve and bone liniment, number six, or cedar oil. I found great relief from cedar oil. Perhaps it would be better to bathe thoroughly with some of these or other liniments first, and apply the poultices afterwards, and continue both applications at the same time, changing the poultices before they become cold.

If the pains are not thus stopped and the gathering driven back, but suppuration is to take place, the next best thing is to hurry it along, so that the water or matter that has gathered next the bone may be removed before the bone shall have become affected thereby; and for this purpose, provided the pain has become so seated that we could tell the precise spot of its location, I would, were I the sufferer, have it lanced to the bone, or lanced in part and then burnt down to the bone with lunar caustic. If this is not done, and suppuration takes place, and a drawing salve be required, take one part of Castile soap, one part beef's gall, one part rosin, two parts beeswax, and enough of white pine turpentine, honey, or flaxseed oil to make them into a salve; melt all together and apply it.

Should it not be inclined to heal, and there appears to be dead flesh in the hole, take the green or fresh root of the May-apple (mandrake), of the proper size, and after preparing it nicely, insert it in the orifice to the bottom, and keep it there from twelve to twenty-four hours; then take it out, and wash out the orifice with Castile soap-suds, warm (using a syringe for that purpose, if necessary); then insert another piece of the May-apple root, and continue this course as long as the patient can reasonably bear it, or until it becomes raw, or bleeds, and live flesh appears. This will make it run or suppurate profusely, and it must be washed often with warm soap-suds, and the salve must be continued. If dead flesh should again accumulate, use the May-apple root as before directed. Should proud flesh appear, use burnt alum, pulverized, on it. Care must be taken not to let it heal at the surface before it is healed at the bottom.

Should the bone appear to be affected, or should the orifice not readily heal at the bottom, make a powder of soft soap and fine salt, in equal parts, mixed together and stewed over a slow fire until reduced to a fine powder by stirring occasionally, and put a little of it as near to the bone or the bottom of the orifice as possible. For this purpose use a hollow tube (a piece of elder or goose-quill will sometimes answer) by putting into it as much of this powder as you wish to insert at one time; then put in a rod or piston, push the powder close to the end you wish to insert, and hold it there; then insert the tube, with the rod and powder in it, as far as possible into the orifice, pull the tube out, or partly out, holding the rod to its place, so as to force the powder out of the tube and as near to the bottom of the hole as possible: repeat this occasionally, and it will eat and cleanse out all the rotten substance.

This powder, the May-apple root, with frequent washing with warm Castile soap-suds, and the above prescribed salve, or the following one, which is known to be better for healing purposes, will permanently heal, I think, any white-swelling ulcer: Take of beech bark, dogwood bark, sassafras bark, sumach bark, elder (common) bark, elder (box) bark, older bark and spikenard root, of each half a pound, and quarter of a pound of hyssop. Boil all these together in a gallon and a half of water down to one quart; strain, and add half a pound of mutton tallow, one ounce of rosin, and one tablespoonful of honey; then simmer down to a salve, and apply it twice a day until cured. This is known to be good.

SALVE FOR ULCER SORES.—Take half a pint of honey, simmer it over a slow fire, and skim off the froth; then take a lump of blue vitriol the size of a common bean, and dragon's blood half the size of the vitriol; pulverize, mix them in the honey while it is hot, and when it becomes cold put a little on the sore at a time. This is an excellent remedy for old sores on the legs, etc.

TO STOP BLEEDING FROM WOUNDS.—Bleeding from a

wound on man or beast may be stopped by a mixture of wheat flour and common salt, in equal parts, bound on with a cloth. If the bleeding be profuse, use a large quantity—say from one to three pints. It may be left on for hours, or even days, if necessary. In this manner the life of a horse was saved which was bleeding from a wounded artery. The bleeding ceased in five minutes after the application. It was left on for three days, when it worked loose, and was easily removed from the wound, which very soon healed.

Another Remedy.—Take of brandy or common spirits two ounces, Castile soap two drachms, and potash one drachm; scrape the soap fine, and dissolve it in the brandy, then add the potash; mix well, and keep in a close phial. When applied let it be warmed, and dip pledgets of lint, and the blood will immediately congeal. It operates by coagulating the blood a considerable way within the vessel. A few applications may be necessary for deep wounds, and where limbs are cut off.

Another Remedy.—Take the fine dust of tea, or the scrapings of the inside of tanned leather, and bind it close upon the wound, and the blood will soon cease to flow. These articles are at all times accessible, and easy to be obtained.

To Stop Bleeding from the Nose.—Extraordinary as it may appear, a piece of brown paper, folded and placed between the upper lip and the gum, will stop bleeding at the nose. Try it.

Another Remedy.—Mr. Negrier states that the hæmorrhage may be almost instantaneously checked by raising the arm on the same side as that of the nostril from which the blood flows.

Another Remedy.—Take the common netch-roots, dry them, carry them in the pocket, and chew them every day. Continue this for three weeks.

Another Remedy.—A piece of ice laid on the wrist will often stop bleeding at the nose.

Another Remedy.—Take some smoked beef, grate it fine, stuff the nostril full, and let it remain till the bleeding stops.

Another Remedy.—Take of blue vitriol and alum each an ounce and a half, and of water one pint; boil them until the salts are dissolved, then filter the liquor, and add to it one drachm of the oil of vitriol. This is used to stop bleeding from the nose or other parts. Wet dossils or cloths with it, and apply them to the part.

To Stop Bleeding from the Lungs or Stomach.—Take half a pound of yellow-dock root, dry it thoroughly, pound it fine, boil it in a quart of sweet milk, strain it off, and drink a gill three times a day. Take also a pill of white pine turpentine every day, to heal the vessels that leak.

Another Remedy.—Take a handful of blood-weed—it is about waist or shoulder high, one stalk from the bottom, has a very bushy top when it is green, grows in old fields, and is called by some horse-tail or white-top—pound it and press out the juice; give the patient a tablespoonful at a time, once an hour, until the bleeding stops. If it be dry, boil it strong, and give the tea very strong, three or four spoonsful at a time.

Another Remedy.—Take a teaspoonful of loaf sugar and rosin, in equal parts, powdered and mixed, three or four times a day, and eat freely of raw table salt.

Teas and other Drinks for the Sick.—*To make Beef Tea for the Sick.*—Take a pound of entirely lean beef, and cut it into small pieces; put it into a gallon of water, with a piece as large as one's hand of the under crust of a loaf of wheat bread, and a little salt; let the whole boil till it is reduced to two quarts, and strain, when it is fit for use. This is for a patient not very weak. For a very weak patient, take the beef as above, pour boiling water on it, cover it up, and let it stand until cold; then strain it off, and warm it as the patient requires, seasoning it a little with salt.

Beef Tea.—This is made best by cutting up tender, juicy beef in bits about an inch square, put into a strong bottle, cork it tightly, and set it in a kettle of cold water. Boil it about two hours. The fluid thus obtained will be the pure nutriment of the meat, and its tonic effects are powerful. Physicians consider it better than alcoholic stimulants in cases of extreme exhaustion, where there is a feverish tendency in the patient.

Tamarinds.—Boil two ounces of the pulp of tamarinds in two pints of milk, and then strain. Use it as a refrigerant drink.

Another.—Dissolve two ounces of the pulp of tamarinds in two pints of warm water, and allow it to get cold. Use as a refrigerant.

An Excellent Drink.—Toast ripe Indian corn quite brown, or even a little black, steep it in hot water, and drink when cold. This is one of the best drinks for the sick, and often arrests sickness of the stomach when other remedies fail.

A Good Drink.—Gum arabic, tapioca, oatmeal, rice, slippery elm, and ripe baked apples, are all good for the very sick, delicate and feeble. Corn-starch pudding, without eggs, is also good.

Milk Porridge.—Beat a little flour into a paste, then stir it into a quart of boiling milk, and cook it well.

Meal Gruel.—Stir a little Indian meal or oatmeal into water, and boil thoroughly.

To PURIFY ROOMS.—*Cholorine Gas.*—The following is said to be one of the most powerful and efficacious disinfectants known. It was used successfully throughout Great Britain and Ireland some years ago, when cholera was so prevalent and fatal there: One part of black oxide of maguese, three parts of common salt, and pour over them a little common vitriol. This makes the gas, a light-colored smoke. Do not inhale it, but place it on a table in the hall of the house; the fumes will then get up stairs and purify the whole building. A pound will purify the house for a month. It is also a good preventive of typhus fever, even in the worst localities, and it is said that cholera never comes where this is freely used.

Powder for Fumigating Sick Rooms and destroying Contagion. Take cascarilla (reduced to a coarse powder), chamomile flowers, and anisseed, equal parts of each, say two ounces; put some hot cinders on a shovel, sprinkle this gradually on it, and fumigate the chambers of the sick. It takes away all smell, and keeps off infection.

Worth Knowing. — Green copperas dissolved in water will effectually concentrate and destroy the foulest smells, and if placed under beds in hospitals and sick rooms, will render the atmosphere free and pure. For butchers' stalls, fish markets, sinks, and wherever there are offensive or putrid gases, dissolved copperas sprinkled about will in a day or two purify the atmosphere, and an application once a week will keep it sweet and healthy.

To Purify Rooms. — Dissolve a few spoonsful of chloride of lime in a saucer, and place it in the apartment.

CURE FOR THE ITCH. — Take half a pound of hog's lard, four ounces of spirits of turpentine, two ounces of flour of sulphur, and mix them together cold. Apply it to the joints and rub it in the palm of the hands. If there are any raw spots, apply a little for three nights, when going to bed. It is an infallible cure.

CURE FOR COLIC.—*Bilious Colic.*—"The following recipe," the "Mobile Tribune" says, "has been handed to us, and we are assured that it is a certain remedy for that distressing disease, as it has never been known to fail in a single case: Take quarter of a pound of plug chewing tobacco, tear it well to pieces, put it into a vessel, and pour upon it enough boiling water to moisten and swell the leaves, then lay it on a cloth and apply to the seat of pain."

Painters' Colic.—It is a fact not generally known, that what is called lemon syrup, made from sulphuric acid, is an effectual preventive of this disease. Those who labor in white-lead manufactories ought never to be without it, for where it has been used the disease is unknown: so says the celebrated German chemist, Liebeg.

Wind Colic in Women and Children.—Take equal parts of ginseng and white-root, half as much calamus or angelica seed, dry them, pound very fine, and mix them together. Dose—a teaspoonful for a grown person; for children less, according to their age. Repeat the dose every half hour, if required. It rarely, if ever, fails.

Another Remedy.—It is said that a pill of asafœtida will sometimes relieve the pain when other remedies fail.

For Colic or Cramp in the Stomach.—Take ten drops of the oil of lavender, on sugar or in wine. Repeat the dose once an hour, if required. Or, drink freely of composition powders, or "number six." Or, a teaspoonful of peppermint, in half a glass or less of brandy, with ten drops or less of laudanum.

SORE NIPPLES AND BREASTS.—When the infant stops sucking, apply a plaster of balsam of fir. It will cure in three or four days.

For Sore Nipples.—Pour boiling water on nutgalls (or oak bark, if galls cannot be obtained), and when cold strain it off and bathe the parts with it, or dip a cloth in the tea and apply it; or twenty grains of tannin may be dissolved in an ounce of water and applied. This application of a few drops of collodeon to the raw surface has been very highly recommended by some physicians. It forms, when dry, a perfect coating over the diseased surface.

Another Cure.—Take the kernels of hickory nuts that are a year or two old, and press out the oil, which can be done by placing them between two flat-irons, and rub the nipples with the oil.

Salve for Women's Sore Breasts. — Take one pound of tobacco, one pound of spikenard, half a pound of comfry, and boil them in three quarts of chamber-lye till almost dry; squeeze out the juice, add to it pitch and beeswax, and simmer it over a moderate heat to the consistence of salve. Apply it to the part affected.

Cake in the Breasts and Sore Nipples. — It frequently happens, when the milk first begins to secrete after child-birth, that the breasts become much swollen and caked, and sometimes suppurate, or gather and break, or have to be lanced. Nothing can be more painful or cause more suffering. All this may easily be avoided, it is said, by using the Mexican mustang liniment freely.

For Hard Breasts. — Melt fresh butter, and while warm rub it on the breasts, and cover them with flannels to keep them warm. This is good to soften the breasts and make the milk flow freely.

Another Remedy. — Take turnips that are roasted soft, mashed, and mixed with sweet oil, and apply them to the breasts twice a day, keeping them warm with flannels.

SORE AND WEAK EYES. — White vitriol one teaspoonful, sugar of lead one teaspoonful, gunpowder two teaspoonsful, to one quart of rain water, mixed and shaken well together. Wash the eyes three times a day with this mixture. This is an infallible cure.

Another Cure. — Sulphate of zinc ten grains, sugar of lead twenty grains, rose water one pint; dissolve each separately and mix. Turn off the clear liquor for use.

Wash for Inflamed Eyes. — A weak solution of common salt, applied to the eyes two or three times a day, is said to be very good.

To take a Film off the Eyes. — Take sugar of lead, make it very fine; take an oat straw, cut it short, so as to be hollow through, dip the end of the straw in the powder, and blow a little of the sugar of lead into the film morning and night.

When the film is almost consumed, apply to it a drop of hen's fat once a day until it is well.

For Weak Eyes.—Take three ounces of rectified spirits of wine, two pugils of elder flowers, and half a drachm of gum camphor; mix them well, and bathe *around* the eyes and on the temples, night and morning. This is known to be an excellent remedy for weak eyes.

ASTHMA, PHTHISIC AND QUINSY. — *Asthma.* — Take two ounces of spikenard root, two ounces of sweet flag, two ounces of elecampane, and two ounces of common chalk; beat very fine in a mortar, add a pound of honey, and beat all well together. Dose—a teaspoonful three times a day.

Another Cure. — The juice of radishes is good for asthma. A small dose of castor oil occasionally will be found beneficial, or new milk every morning and evening. Other remedies are recommended, such as smoking, inhaling hot vapors, garlic, etc.

Another Cure. — The following is also recommended for this complaint: Two ounces of the best honey and one ounce of castor oil, mixed; a teaspoonful to be taken every night and morning.

Phthisic.—For this complaint in children take four ounces of seneca snake-root, four ounces of spikenard, four ounces of parsley root, and two ounces of licorice stick; boil them together in four quarts of water; strain, sweeten with loaf sugar or honey, and let the patient drink a small glass night and morning.

For Quinsy.—Sweat the throat with spotted cardis, boiled in milk and water, by holding a pot of it under the throat as hot as can be borne; hold some of it in the mouth, and when the swelling has gone down wear a piece of black silk about the throat constantly, and it will prevent a recurrance of the disease.

Another Cure.—Take half a pint of molasses, a tablespoonful of lard, and a roll of brimstone an inch long; melt over a slow fire, and when cold drench with it. It is said to be almost an infallible cure.

NERVOUS AND SICK HEADACHE.—*Nervous Headache.*—Essence of turpentine is said to be a specific for nervous headaches, even when accompanied with vertigo. Though when applied to the skin it produces redness and irritation, it is perfectly harmless to the most delicate subjects when taken internally.

Sick Headache.—It is said that two teaspoonsful of pulverized charcoal, drank in half a tumbler of water, will in less than fifteen minutes give relief to this disease, when caused (as it is in most cases) by superabundance of acid on the stomach. Lemon juice will often correct acidity of the stomach. Strong boneset tea is also good.

Neuralgic Headache.—The application of towels wrung out of hot water to the forehead and temples, is represented to be a certain and speedy cure for headaches arising from neuralgic affections.

Volatile Liniment.— Two ounces of spirits of ammonia, ten ounces of sweet oil, and one ounce of alcohol; put all into a bottle, shake well, and it is ready for use. It is a remedy for all external bruises where the skin is not broken, for pains in the side, back or limbs. It is excellent for the headache: inhale it gently, and apply sparingly to the temples and back of the neck. There is nothing better for the sting of a bee, frosted feet or chilblains. Wherever pains exist this liniment is excellent, and no family will do without it when its value is known. So says the "Country Gentleman."

Pills for Sick Headache.—Take twenty grains of scammony, twenty grains of Cape aloes, twenty grains of pulverized rhubarb, twenty grains of Castile soap, ten grains of Jamaica ginger, and three grains of tartar emetic; mix, and form into twenty-four pills; take three or four on going to bed, or as soon as the symptoms are felt. Take from six to eight when the pain has been felt from ten to twelve hours. As soon as costiveness begins take a dose of these pills, and you will not be likely to have the sick headache.

TOOTHACHE.—To cure toothache take a piece of sheet zinc about the size of a sixpence, and a piece of silver (say quarter of a dollar), place them together, holding them between the teeth, and contiguous to the defective tooth. In a few minutes the pain will be gone, as if by magic. The zinc and silver acting as a galvanic battery, will produce an effect on the nerves of the tooth sufficient to establish a current, and will consequently relieve the pain.

Another Cure.—Take a spoonful of black pepper, and mix it with the same quantity of salt; then place the same mixture upon a shovel, hold it over the fire until it smokes and then inhale some of the smoke through the nose.

Another Cure.—Wet enough of cotton to fill the cavity in the tooth, and cover it with alum and salt, in equal parts, finely pulverized, and apply it.

Another Cure, if the tooth be hollow.— Take gum camphor, gum opium, and spirits of turpentine, in equal parts; rub them in a mortar to a paste, dip lint in the paste and put it into the hollow of the tooth every time after eating. Do this for three or four days, and it will entirely cure the tooth from ever aching.

EARACHE.— Take a small piece of cotton batting or raw cotton, make a depression in the center with the finger, and fill it with as much ground pepper as will rest on a five-cent piece ; gather it into a ball and tie it up, dip the ball into sweet oil, and insert it in the ear, covering the latter with cotton wool, and use a bandage or cap, to retain it in its place. Almost instantaneous relief will be experienced, and the application is so gentle that an infant will not be injured.

Another Cure.—Cotton wool, wet with sweet oil and paregoric, and put into the ear, is very good.

Another Cure.—A poultice of roasted onions, mashed, with a few drops of laudanum added, is also very good to put into the ear.

BURNS AND SCALDS.—The best thing for a burn is the following, and every family ought to know it: As soon as possible throw a little green tea into hot water, and let it steep; stir an Indian-meal poultice, spread the tea-leaves on the poultice, and put it upon the burn or scald, whatever it may be. If burnt with gunpowder, it will take it out, and the skin will be as clear as ever.

Another Cure.—Cotton batting, moistened with linseed or sweet oil, and applied directly to the burn, is very efficacious; the linseed oil is preferable, as it allays the pain, while at the same time it extracts the fire. The cotton should not be removed when the skin is off, until the burn is healed, as the new skin will adhere to it while forming, and come off with it. If the burn is washed in lime water, previous to applying the cotton, it will not be so liable to leave a scar.

Another Cure.—Honey and molasses spread on a cloth and applied twice a day is good.

Another Cure.—The best application for a burn or scald that I have ever seen used is wheat flour, put on dry. Put on enough to exclude the air, bind it on, and let it remain on till it comes off itself.

PILES.—A gentleman has handed in the following recipe, as a certain cure for this complaint. It has often been tried and always proved successful: "One scruple of powdered opium, two scruples of flour of sulphur, and one ounce of simple cerate. Keep the affected parts well anointed. Be prudent in your diet, do not overload the stomach, exercise as much as possible in the pure air, and keep your mind cheerful by thanking God for the manifold blessings which you enjoy."

Ointment for the Piles. — Take four ounces of hog's lard and half an ounce of laudanum; mix, and apply every night at bed time.

Another Remedy. — To one bottle of Mexican mustang liniment add one drachm of iodine and two ounces of lard; rub them well together in a mortar, and anoint the parts affected every morning and evening, after washing clean with Castile soap and cold water. If this treatment is closely observed, a cure is warranted, no matter how bad, or of how long standing they are.

STIFF JOINTS AND CONTRACTED SINEWS. — Take half a pound of hog's lard, put into it a small handful of melilot (or meliot) green, stew them well together, strain off, add one ounce of rattlesnake's grease, one ounce of olive oil, and ten drops of oil of lavender, mixed well together. Use three times a day, rubbing in well with the hand.

For Shrunken Sinews or Stiff Joints. — Take half an ounce of yellow basilicon, half an ounce of green melilot, half an ounce of oil of amber, and a piece of blue vitriol the size of a chestnut; simmer them together to a salve or ointment, apply to the part affected and on the joint above; repeat it, and it will perform the cure.

Another Remedy. — Take fresh butter and fish-worms, stew them together, and while warm squeeze out the oil, with which bathe the parts well and often. This is very good.

OLD SORES AND ULCERS. — Scrape yellow carrots, wilt them on a pan or fire-shovel, very soft. It removes both the inflammation and swelling, and is an excellent poultice for an ichorous breast.

Healing Salve for Ulcers. — Melt and mix together three parts of beeswax, two of lard, one of mutton suet, and one of common rosin. Ulcers of long standing, when irritable or painful, with a burning sensation, or when they do not show a disposition to heal, or when the edges are covered with a dead, white skin or scurf, should be penciled on the edges every eight or ten days with lunar caustic, and then poulticed, until the inflammation subsides, after which they may be dressed with the salve.

Foot's Ointment.—Take one pound of hog's lard, one pound of mutton tallow, half a pound of oil of spike, and heat them over a moderate fire until they are united, then add as much beeswax and rosin as will make it a salve. This is the renowned "Foot's Ointment," and cures all common sores, where there is no inflammation.

Inward Hurts and Ulcers. — Two ounces of sassafras root bark, two ounces of coltsfoot root, two ounces of blood root, one ounce of gum myrrh, one ounce of winter bark, and one ounce of suckatrine aloes, steeped in two quarts of spirits. Drink a small glass every morning, and live on simple diet as much as possible. For a constant drink make a beer of one peck of barley malt, two pounds of spikenard root, one pound of comfry root, two pounds of burdock root, five pounds of black spruce boughs, one pound of angelica root, and quarter of a pound of fennel seed, for ten gallons of beer. Drink one quart a day, and let your exercise be light.

An Excellent Medicine.—Take of elecampane, comfry, masterwort, spikenard, angelica and ginseng roots, of each one pound, boughs of fir two pounds, and chamomile one pound; put them into a still, with a gallon of rum and two gallons of water, draw off six quarts, and drink a small glass night and morning.

Another Excellent Remedy. — Good for all sorts of inward weakness, inward fevers, coughs, or pain in the side, stomach or breast: Take twenty pounds of fir boughs, one pound of spikenard, four pounds of red clover; put them into a still, with ten gallons of cider, draw off three gallons, drink half a gill night and morning.

SWELLINGS.—*Red Salve for Swellings in formation.*—Take a pound of linseed oil, half a pound of sweet oil or fresh butter, one pound of red lead; boil them together, stir while boiling, then lessen the heat and add to it two pounds of beeswax and one pound of rosin, and stir them together till cold.

Another Remedy.—Nothing is so good to take down swellings as a soft poultice of stewed white beans applied in a thin muslin bag, and renewed every hour or two.

Another Remedy.—Take spirits of turpentine, number six, flaxseed oil, and oil of spike, in equal parts; mix them well together, and bathe well the part affected every night and morning. This is good also for rheumatism. A decoction of smartweed and vinegar, as a fomentation, is also very good to reduce a swelling.

SPRAINS.—By sprain is meant the partial displacement or twisting of a joint. The ankle and wrist joints are most liable to this accident. The best thing for a sprain is, first to foment the parts well for an hour or two with rags and warm water; then mix equal parts of sugar of lead and opium in one quart of boiling water, and apply continually with rags. Absolute rest is necessary.

Another Remedy.—An old friend has handed in the following recipe for sprains, the publication of which, if it be as good as he deems it, may be productive of benefit to some of our readers: One pint of soft soap (country preferred), one pint of whisky; mix and boil them until the quantity is reduced to about one pint: it will then be of the consistency of a plaster salve. Take a thick cloth, linen or cotton, sufficiently large to cover the parts injured, spread the mixture, and sprinkle the same with black pepper. Apply this and let it remain for forty-eight hours. In nine cases out of ten a cure will be effected.

DRAWING SALVE.—One part of beef's gall, one part of

Castile soap, one part of rosin, and two parts of beeswax; pulverize and mix them with enough of white pine turpentine, honey, flaxseed oil, or spirits of turpentine, to make a salve. Simmer all together, and apply it.

DROPSY. — A correspondent of a respectable newspaper recommends the following as an excellent remedy for water on the chest: Take quarter of a pound of dried milk-weed, cut small, pour on it a quart of boiling water, simmer to one pint, when cool add a pint of the best Holland gin, pour both liquor and roots into a decanter, cork it tight, and let it stand for twelve hours. Dose — for an adult, half a wine glass every three hours, day and night. If it nauseates too much, the dose may be varied. Effect seen in three or five days.

Another Remedy.—Take half a pound of blue flag root and half a pound of elecampane root, boil in two gallons of fair water to one quart, and sweeten with one pint of molasses; let the patient take half a gill three times a day, before eating.

Another Remedy.—Take one pound of sassafras bark of the root, one pound of prickly ash bark, half a pound of spice wood bush, three ounces of garlic, four ounces of parsley roots, four ounces of horse-radish roots, four ounces of black birch bark; boil all in three gallons of malt beer, and drink a gill three times a day.

Another Remedy.—Take two gallons of good vinegar, three pounds of rusty iron, one handful of garlic roots and tops, a handful of horse-radish roots, and two handsful of grape-vine ashes; put all together, simmer down to one gallon, and take a large drink three or four times a day. Eat light diet only.

WORMS IN CHILDREN.—There are many things helpful to children troubled with worms. The bark of witch hazel or spotted alder is one. Shave the bark down toward the root, steep in a pewter vessel, do not let it boil, but steep it on a moderate heat very strong. A child a year old can take a tablespoonful, and more according to age. Let them take it four or five times a day for several days. It is sure and safe.

Another Cure.—Take sage, powder it fine, and mix it with honey. A teaspoonful is a dose.

Another Cure.—Sweetened milk, with a little alum added to it, is very good to turn worms.

Another Cure.—Flour of sulphur mixed with honey is also very good.

Another Cure.—Take a piece of steel, heat it very hot in a smith's fire, then lay on it a roll of brimstone, melt the steel, let it fall into water, it will be in round lumps, pound them very fine, mix the dust with molasses, and let the child take half a teaspoonful night and morning, fasting.

Another Cure.—Wild mandrake roots, dried and powdered, and mixed with honey. Give to a child a year old as much of the powder as will lie on a sixpence; to be taken in the morning, fasting, three or four times successively.

Another Cure.—If a child is taken with fits by reason of worms, give as much paregoric as the child can bear. This will turn the worms and ease the child.

Another Cure.—Salt and water is good to turn worms, and by giving a dose or two of flour of sulphur, mixed with molasses or honey, afterwards, it brings off the worms without anything else.

Another Cure.—From one-half to one teaspoonful of liquor, into which a little garlic has been infused, given every morning for a week, is said to be good for children troubled with worms.

A Preventive.—To prevent children from having worms, let them eat onions, raw or cooked—raw are best.

GRAVEL AND STOPPAGE OF WATER. — *For Gravel in Bladder or Kidneys.*—Make a strong tea of the herb called heartsease, and drink plentifully. Or, take the root of Jacob's ladder, and make a very strong tea, and drink plenty of it. It is a most certain remedy. Jacob's ladder is a vine that grows often in rich, interval soil, near a wood or brush that stands near grass land. It comes up with one stalk about breast high, single, then runs off into a number of branches covered with green leaves, and the fruit is a large bunch of black berries. When ripe the bunch hangs down under the leaves by a small stem.

For Stoppage of Water. — Take a spoonful of honey bees, and as much buds of currant bushes; steep them in hot water, very strong, and drink two spoonsful every half hour.

HICCOUGH. — A contemporary publishes the following cure for hiccough: "You may be the medium of relief to many who suffer from hiccough if you will state that it can be cured immediately merely by placing the hand of any person upon the pit of the stomach of the one afflicted, and a persistent stare of the sufferer into that of the person who undertakes the cure. I do not recollect how I first learned this means of cure, but for twenty years I have known it to be a fact, without a single failure. I will not attempt to philosophize upon it, or give a theory. All that I know is that I have relieved many, and have been cured hundreds of times as above stated."

RHEUMATISM. — Bathe the parts affected with water in which potatoes have been boiled; apply as hot as can be borne just before going to bed, and by the next morning the pain will be much relieved, if not removed. One application of this simple remedy has cured the most obstinate rheumatic pains. Several persons have recently testified to the value of this remedy.

Another Cure. — Take a handful of prince of pine, a handful of horse-radish root, of elecampane root, prickly ash bark, bittersweet root bark, wild cherry bark, mustard seed, and a pint of tar water put into two quarts of brandy; drink a small glass every morning, noon, and night before eating. Bathe the part affected with salt and rum, by a warm fire.

For Rheumatism in the Loins. — The oil of sassafras used internally and externally. Ten drops on loaf sugar is a dose. Anoint the part affected with the same, and repeat as often as needful. Drink poke berries in brandy for three weeks every day. Or, drink brandy, and bathe the part affected with salt and rum, as hot as can be borne, by a fire. Repeat it for six days.

For Rheumatism and Gout. — A medical correspondent of the "York [England] Courant" says the advantages of asparagus are not sufficiently estimated by those who suffer from rheumatism and gout. Slight cases of rheumatism are cured in a day by feeding on this delicious esculent, and more chronic cases are much relieved, especially if all acids are carefully avoided, either in food or as a beverage. The Jerusalem artichoke has also a similar effect in relieving the disease. The heads may be eaten in the usual way; but tea made from the leaves and stalks, and drank three or four times a day, is a sure remedy, though not equally agreeable.

Another Cure. — Lemon juice is now used quite extensively by physicians in London as a cure for rheumatism. Three tablespoonsful a day for adults is the average dose.

Another Cure. — Take about half a gallon of pipsissewa (wintergreen), bruise and put it into a stone jug that holds one gallon and a half, and pour on it one gallon of pure rye whisky; shake it up occasionally for two or three days, and then drink a wine glass of it morning, noon and night. Use a little cream of tartar once a day, dissolved in cold water.

Liniment for Rheumatism. — One pint of alcohol, two ounces of sulphate of ether, two ounces of ammonia, one ounce of tincture of aconite, one ounce of gum camphor, one-fourth of an ounce of spirits of turpentine; mix, and bathe the part affected two or three times a day.

NUMB PALSY.— When a person is taken with the numb palsy, bleed freely, if possible, give a tablespoonful of flour of sulphur once an hour, bathe the part affected with spirits of hartshorn, take one pound of roll brimstone, boil it in four quarts of water to one quart, and let the patient drink a tablespoonful every hour. If applied early it will finally carry it off.

POLIPUS.— Take two ounces of blood-root, dry it, pound it fine, then add quarter of an ounce of calix cinnamon, two ounces of scoke-root, snuff it up the nose and it will kill the polipus; then take a pair of forceps and pull it out. Use the snuff until it is cured. If the nose is so stopped that it cannot be snuffed, boil the same and gargle it in the throat, and sweat the head with the hot liquor until it withers so as to use the snuff.

SCARLET FEVER.—The following recipe, says the "Boston Star," is given as almost a specific for this disease. The gentleman who gave it mentions numerous violent cases in which he employed it, in all of which the cure was almost immediate—within one day in every instance:

Immediately on the first symptoms, which is sore throat, give a full dose of jalap—to an adult, sixty, seventy, or even eighty grains; at night give strong red pepper tea, from a teacupful to a pint, according to age and the violence of the symptoms; the next day give a small dose of jalap, say half the quantity given the day before, and continue the pepper tea at night; on the third day, if there is any soreness in the throat, give a dose of salts, which will generally effect a cure. The doses must of course be regulated according to the age of the patient.

Supposed Preventive.—A writer in the "Boston Post" says with great confidence that one globule of belladonna taken every morning by each member of the family—adults, children, servants, and all inmates — will certainly prevent the spread of the disease in every household that may adopt it, as certainly as vaccination will prevent the small-pox. Ten cents will purchase a year's supply.

SCARLATINA AND MEASLES. — Mr. Witt, a member of the Royal College of Surgeons, has published a pamphlet, in which he states that carbonate of ammonia is a specific for the cure of scarlet fever and measles. He cites Doctor Pearl, of Liverpool, and other practitioners, who have never lost a case, out of hundreds, since adopting this remedy. Two drachms of the bicarbonate of ammonia are dissolved in five ounces of water, and two tablespoonsful of the solution given every two, three, or four hours, according to the urgency of the symptoms. No acid drink must be taken, but only water, or toast and water. The system is to be moved by a dose of calomel, if necessary. The room must be well ventilated, but the patient protected from the slightest cold or draught. Gargles should also be employed for cleaning the throat. The ammonia, it is said, counteracts the poison which causes scarlatina, and also acts upon the system by diminishing the frequency, and at the same time increasing the strength of the pulse. As so many children die from these diseases in this country, this remedy ought to receive a fair trial from the profession.

Another Cure.—Rubbing the body all over with fat bacon, raw, three times a day, as soon as the fever is manifest, is said to be a good remedy. A poultice of raw cranberries, pounded fine, is good to reduce the inflammation.

AGUE, CHILLS AND FEVER. — *Cure for Chills and Fever.*— As soon as the chill appears take a dose of calomel, and after it operates two or three times take a small dose of castor oil. The next morning commence taking the following: one ounce of pulverized Peruvian bark, half an ounce of cream of tartar, quarter of an ounce of cloves, which must be ground fine; mix all together, and put into a bottle, stir about half a pint of wine, brandy or whisky into it, and shake the bottle well before using. Take a small wine glass three times a day before meals; for children half that quantity at a dose.

Another Remedy.— Take one pint of brandy or whisky, one gill of the tincture of blood-root, half a teaspoonsful of oil of pennyroyal, half a teaspoonful of oil of sassafras. half a teaspoonful of oil of peppermint, and forty grains of quinine; mix all together in a bottle, and it is ready for use. Commence using the medicine four hours before the chill comes on, and for an adult give one tablespoonful every hour for four hours. If this does not stop the chill the first day, repeat the dose the second day, and so on till cured.

Another Remedy.—It is said that a tea made of green corn blades, and drank occasionally when the chills are off, will prevent their return. When the chills are once broken, if they return at all, they usually occur periodically, at the end of the first or second week. To prevent this take a dose or two of the medicine on the sixth and seventh days from the time they were broken, and repeat the same at the end of the second and third weeks.

Another Cure.— Take two ounces of Peruvian bark, two ounces of wild cherry tree bark, one ounce of poplar bark, one ounce of prickly ash bark, one ounce of Virginia snake root, and about half a tablespoonful each of cinnamon bark, cloves and ginger—all finely pulverized, and put into half a gallon of good port wine, or good cherry or French brandy, and let it stand two or three days before using. Dose — a wine glass of it three or four times a day, and a cure will soon be effected. This is much better and safer than quinine, and the cure much more effectual.

Ague in the Face.— Apply a poultice made of flour and ginger. A mustard poultice is also good, but it is apt to scar the face. Hops, steeped and applied to the face, often afford relief.

Ague Medicine.— Rhubarb, columba, and essence of peppermint, each one ounce, one pint of water, and forty-five grains of quinine. A tablespoonful once an hour until it operates as physic, then the same amount three times a day. To keep it, add one gill of whisky.

Fever and Ague Mixture. — One pint of the best brandy, dissolve in it one ounce of camphor, add half an ounce each of cloves and jalap, two ounces of Peruvian bark, one ounce of Virginia snake-root, and one pint of water; boil cloves and root with the water, to one half; strain, and mix the others in powder with the above. Dose — a tablespoonful three times a day, in the absence of the fever.

Fever and Ague Pills. — Two ounces of aloes, one ounce each of gamboge and cream of tartar, and half an ounce of saltpetre; divide into pills of five grains each, of which three are a dose. Powder and sift the whole, and mix in a mass with syrup of molasses.

For Dumb Ague, Weakness, and General Debility.—Take two drachms each of pulverized Peruvian bark, pulverized gentian root, quassia chips, and orange peel; put all into one quart of rye whisky. Dose — for an adult, one tablespoonful, diluted with water, ten or fifteen minutes before each meal.

INFLUENZA. — Boil strong vinegar and inhale the steam through a clean pipe or funnel, and drink freely of boneset or hoarhound tea, and bathe the feet in warm lye water.

BRONCHITIS.—Take green mullen leaves, dried, and inhale the smoke thereof through a new or clean pipe, letting the smoke pass out at the nose.

Another Cure.—Honey diluted with water, and used as a gargle, is good.

HOARSENESS.—One drachm of freshly scraped horse-radish root, to be infused with four ounces of water in a close vessel for two hours, and made into a syrup, with double its weight in vinegar, is an improved remedy for hoarseness. A teaspoonful, it is said, has never been known to fail in removing it.

Another Remedy. — Take half a pint of good vinegar, a little ginger, horse-radish, butter and honey or sugar, and simmer together. Dose — from one to two teaspoonsful occasionally.

FROZEN LIMBS. — When any part of the body is frozen, it should be kept perfectly quiet till it is thawed out, which should be done as promptly as possible. As freezing takes place from the surface inwardly, so thawing should be in the reverse order, from the inside outwardly. The thawing out of a portion of flesh, without at the same time putting the blood from the heart into circulation through it, produces mortification; but by keeping the more external parts still congealed till the internal heat and the external blood gradually soften the more interior parts, and produce circulation of the blood as fast as thawing takes place, most of these dangers are obviated.

Raw cotton and castor oil have restored frost-bitten limbs when amputation was thought to be necessary to preserve life. It is said to be an infallible cure.

INFLAMMATION OF THE BRAIN. — A kind of cushion of crushed ice put on the entire scalp, is good to prevent this inflammation of the brain, and to arrest convulsions induced by too much blood there.

The same applied to the nape of the neck is good to prevent a pressure of blood on the brain.

The same, or ice-water, on the head, the warm bath, and then mustard plasters on the feet, are good to prevent convulsions.

For Inflammation in the Head. — Take red beets, pound them fine, press out some of the juice, let the patient snuff some of it up into the head, and make a poultice of the beets, and lay it on the head. Keep the bowels open, and strong mustard draughts to the feet.

INFLAMMATION OF THE STOMACH.—A small lump of saltpetre (nitre) the size of a rye grain, dissolved in a tablespoonful of water, and drank, has stopped the vomiting caused by inflammation of the stomach, when other remedies have failed. The patient should then use a little magnesia dissolved in water, for some days, and be very careful in the diet: *let it be light and cooling.* In this disease keep the bowels open with injections. It may become necessary to put a large mustard plaster over the stomach, and when it is removed keep warm flannels over the stomach, bathe the feet in warm lye water, and keep them warm.

CATARRH IN THE HEAD.—Take one part of saltpetre and two parts of loaf sugar, pulverize and mix them together, and snuff a little up the nose several times a day.

SCROFULA.—Common salt dissolved in good brandy is good to disperse scrofulous swellings. Wash and bathe the part affected. It is also good for sore throat and all kinds of ulcers. A teaspoonful of the tincture of gum guaiacum two or three times a day is good for scrofulous affections.

Ointment for the same.— Take one ounce of tobacco, four ounces of white ash moss, four ounces of soot, four ounces of hog's lard, four ounces of tar, and two ounces of antispasmodic drops. Boil the tobacco, moss and soot in two gallons of water down to one gallon, then strain and boil down to a quart, then add the lard and tar, and simmer down to a pint and a half, then add the anti-spasmodic drops, stir until cool, and then apply it. Drink alteratives three times a day, and keep the bowels moderately open with cream of tartar and flour of sulphur, until a cure is effected.

For Diseases of the Skin.—Use the above prescription, and

if the skin is spotted, as in leprosy, apply the tincture of iodine to the spots occasionally.

Another Cure for Scrofula.—The late Nicholas Longworth, of Cincinnati, in a letter to a gentleman in Dayton, gave the following recipe for the cure of scrofula:

"To one tablespoonful of aquafortis two of strong, clear vinegar is added, and two copper cents are dropped in, which creates a strong effervescence, and are left in until it ceases to sparkle. The preparation is twice per day applied to the sore with a soft rag."

Another Cure.—The scrofula, or king's evil, may be cured with a plant called king's evil weed. It grows in wild shady land, under almost all kinds of timber, in the form of the plantain, but the leaves are smaller, and are spotted green and white, a very beautiful plant. When it goes to seed there comes up one stalk in the middle of the plant, six or eight inches high. It bears the seed on the top of the stalk in a small round bud. Take this, root and branch, pound it soft, apply it to the tumor for a poultice or salve, and let the patient drink a tea made of the same for a constant drink. If the tumor is broken open, simmer the root and leaf in sweet oil and mutton tallow; strain it off, and add to it beeswax and rosin, till hard enough for a salve. Wash the sore with liquor made of the herb boiled, and apply the salve, and it will not fail to cure.

LEPROSY.—Put a piece of unslaked lime, the size of a common tea-cup, into a tight vessel, and pour one gallon of water on it; when it cools pour off the water, and wash the spots occasionally with it.

The tincture of iodine is also good to rub on the spots.

A teaspoonful of the tincture of gum guaiacum, taken two or three times a day, or a tea made out of chips of lignum vitæ, and drank often, is good in leprosy and all skin diseases.

DISLOCATION OF THE JAW.—Doctor Buchan says: "The

usual method of reducing a dislocated jaw is to set the patient upon a low stool, so as an assistant may hold the head firm by pressing it against his breast. The operator is then to thrust his two thumbs (being first wrapped up with linen cloths, that they may not slip) as far back into the patient's mouth as he can, while his fingers are applied to the jaw externally. After he has got firm hold of the jaw, he is to press it strongly downward and backward, by which means the elapsed heads of the jaw may be easily pushed into their former cavities."

PARTIAL DISLOCATION OF THE NECK.—Doctor Buchan says: "To reduce this dislocation the unhappy person should immediately be laid on his back upon the ground, and the operator must place himself behind him, so as to be able to lay hold of his head with both hands, while he makes a resistance by placing his knees against the patient's shoulders. In this posture he must pull the head with considerable force, gently twisting it at the same time, if the face be turned to one side, till he perceives that the joint is replaced, which may be known from the noise which the bones generally make when going in, the patient's beginning to breathe, and the head continuing in its natural posture."

FEVER SORE.—*To Stop a Fever Sore from coming to a head and carrying it away.*—Sweat it with flannel cloths dipped in hot brine. The cloths must be changed as often as they are cold for three hours; then wash in brandy and wrap in flannel, repeating it three or four times.

DIABETES IN CHILDREN.—For those so troubled take two ounces of good red bark and one quart of wine; steep the bark in the wine for fourteen hours. Give the patient, if two or three years old, a tablespoonful; if older, a little more at a time.

Another Cure.—Red beech bark stripped off a green tree; dry it well, pulverize it fine, and use it in the same way.

CHAPPED HANDS.—Take one ounce of bitter almonds, peel them and mash them into a paste, with oil of sweet almonds and the yolk of an egg, adding a little tincture of benzoin, so as to form a thick cream. Now add a few drops of oil of caraway. It is to be rubbed on the hands at night, and a soft kid glove to be worn during the treatment.

Another Remedy.—After washing drop a little honey on the hands and rub them together until the stickiness is entirely removed.

TO EXTRACT SUBSTANCES FROM THE NOSTRILS.—When a child has any substance wedged in its nostrils, press the vacant nostril so as to close it, and apply your lips close to the child's mouth, and blow very hard. This method will generally force the substance out of the nostril.

VEGETABLE POISON.—*To Cure Vegetable Poison, Running Ivy, etc.*—Take rosemary leaves or blossoms, make a tea, and drink night and morning.

Another Cure.—Take wild turnips; if green, pound them and press out the juice; if dry, boil them in fair water, and wash the part affected with the clear liquor. Take part of the liquor, add to it a little saffron and camphor, and drink to cleanse the fluids and guard the stomach.

LOCK-JAW.—When any person has the lock-jaw, give him five grains of Dover's powders, then set him in a tub of water as hot as he can bear it, bathe his head with camphorated spirits, let him sit or stand in the water as long as he can bear it without fainting, and bleed him, if possible. Repeat this three or four times. When out of the water put him in a warm bed, wrapped in flannel.

See recipe for Mortification.

WEN.—Take clean linen rags and burn them on a pewter dish; gather the oil on the pewter with lint, and cover the wen with it twice a day. Continue this for some time, and the wen will drop out without any farther trouble.

Another Cure.—Bathe with liniment, and wear a piece of sheet lead on the wen until it disappears. A sure cure.

DISEASES AND REMEDIES.—Red alder cures swellings and strains; witch hazel cures piles; arrow-root is nutriment for the sick; the sap of black oak cures deafness; prickly ash cures fever and ague; bitter ash cures affections of the lungs; white ash bark cures the bite of the rattlesnake; asparagus root cures diseases of the heart, breast, kidneys and bladder; the balm of Gilead cures general debility of the whole system, and many other diseases; balmony cures the jaundice; Jacob's ladder cures diabetes; garden basil cures chronic headaches, colds, fevers and hysterics; bitterwort cures rheumatism, dropsy and asthmatic cough; blood root (be careful how you use it) expels worms, and cures pulmonary affections, diseases of the chest, liver, etc.

FEEDING CHILDREN. — A prominent physician says: "In my practice I have noticed that those children who become ill and die in the spring and summer have fallen victims to the thoughtlessness of parents, who stuff them with roast and fresh meat, at a season when their stomachs required a vegetable diet, easily digested and equally nutritious. I have saved the lives of more children by recommending farinaceous and vegetable food, than I ever did by dosing them with disagreeable medicines."

UNPLEASANT ODOR OF PERSPIRATION. — The "Scientific American" says: "The unpleasant oder produced by perspiration is frequently the source of vexation to persons who are subject to it. Nothing is simpler than to remove this odor much more effectually than by the application of such onguents and perfumes as are in use. It is only necessary to procure some of the compound spirits of ammonia, and place about two tablespoonsful in a basin of water. Washing the face, hands and arms with this leaves the skin as clean, neat and fresh as one could wish. The wash is perfectly harmless and very cheap. It is recommended on the authority of an experienced physician."

PARALYSIS AND APOPLEXY.—Doctor Chapman, of London, has made a new discovery in the treatment of paralysis and apoplexy. The treatment is briefly described as the application of heat to one part of the spine, and of cold to another part. The "Medical Times and Gazette" narrates several cases where parties afflicted had been given over by their medical attendants, and who had been restored to perfect health by the treatment of Doctor Chapman.

DEAFNESS.—It is said that by mixing sulphuric ether and ammonia, and allowing the mixture to stand for fourteen days, a solution is formed which, if properly applied to the internal ear, will remove in almost every case this hitherto considered incurable affection.

Another Cure.—Take three parts of rabbit oil and one of laudanum; mix, and put from a few drops to one-fourth of a teaspoonful into the ear, night and morning.

GOITRE.—Goitre can usually be scattered by a free application of tincture of iodine.

CORNS.—Take white pine turpentine, spread a plaster, apply it to the corn, and let it stay on till it comes off itself. Repeat this three times; it never fails curing.

Celebrated Three Minutes' Salve.—It has never failed in a single instance: One pound of caustic of potash, four drachms of belladonna, two ounces peroxide of manganese; make into a salve.

Positive Cure.—The strongest acetic acid, applied night and morning, with a camel's hair brush. In one week the corn, whether soft or hard, will disappear.

Another Cure.—A corn may be extracted from the foot by binding on half a raw cranberry, with the cut side of the fruit upon the foot. Old and troublesome corns have been drawn out in this way, in the course of a few nights.

Another Cure.—One teaspoonful of tar, same of saltpetre, same of brown sugar. Warm all together; and apply.

WARTS.—These are common everywhere, and not less so than disagreeable. The best plan to get rid of them is to pare them down as much as possible with a sharp knife, and then touch them with a little nitric acid.

Another Cure.—Make a strong solution with corrosive sublimate, and wet the wart three or four times a day; it never fails curing.

Another Cure.—Dissolve as much common washing soda as the water will take up. Wash the warts with this for a minute or two, and let them dry without wiping. Keep the water in a bottle, and repeat the washing often, and it will take away the largest warts.

Another Cure.—Wet the warts, and apply a little saleratus; repeat this a few times, and it will cure the most obstinate warts.

RINGWORMS.—There are two kinds of ringworm, and every old woman has her cure for it. Solution of sulphate of zinc, copper ore, or a weak solution of nitric acid, and citrine ointment. Touching the part with acetic acid is often useful. Great care should be taken to prevent the rest of the family from taking it. All combs, brushes and towels should be well washed after being used, before being used by another person.

Another Cure.—Apply gunpowder, wet, on retiring at night, and let it remain on the ringworm till morning. The oil obtained from roasting a butternut, applied in the same manner as the gunpowder, is good to remove ringworms.

BOILS AND STYES.—Touch them with spirits of turpentine every six hours. This should be applied to boils and styes in their first stages, to effect a cure.

Another Remedy.—A plaster of honey and wheat flour, or soap and brown sugar, is good for boils.

TEETH.—Washing the teeth with vinegar and a brush will, in a few days, it is said, remove the tartar, thus obviating the necessity for filing or scraping them, which so often injures the enamel. The use of powdered charcoal and tincture of rhatany afterwards is recommended to prevent its formation.

The Best Tooth Wash known to the Profession.—One ounce tincture of gum myrrh, one ounce powdered orris root. Wet the brush with the tincture, then dip into the powder, and apply to the teeth.

To Fasten Loose Teeth.—Gargle the mouth frequently with alum and water.

CRAMP.—Two or three spoonsful of strong lye, made of oak ashes, and mixed with molasses, are recommended as a positive cure for cramp.

CHILBLAINS.—Dissolve one ounce of white copperas in a quart of water, and apply the solution occasionally to the affected parts. It must not be used if the skin is broken, or it will do injury.

Another Cure.—Simply bathe the parts affected in the liquor in which potatoes have been boiled, at as high a temperature as can be borne. On the first appearance of the ailment, indicated by inflammation and irritation, this bath affords almost immediate relief. In the more advanced stages, repetition prevents breaking out, followed by a certain cure, and an occasional adoption will operate against a return, even during the severest frost.

Suffocation from Smoke.—A wet silk handkerchief, tied without folding, over the face, it is said, is a complete security against suffocation from smoke, as it allows free breathing, at the same time excluding the smoke from the lungs.

Spider and Mosquito Bites.—These, it is said, may be cured by rubbing them with the plantain leaf, bruised. The plantain is a weed growing wild in most yards and grass plots.
Another Remedy.—A piece of fresh beef placed in the bedroom at night, or a little of the essence of pennyroyal rubbed on the hands and face at night, will protect the sleeper from the bite of mosquitoes.

Sting of Bees and Wasps.—Chalk, wet with hartshorn, is a remedy; so is likewise table salt kept moist with water; also, a piece of raw onion applied.

Sore Lips.—Dissolve a small lump of white sugar in a large spoonful of rose water—common water may be substituted. Mix it with a couple of large spoonsful of sweet oil, a piece of spermaceti half the size of a butternut. Simmer the whole eight or ten minutes.

Court Plaster.—This should be thoroughly soaked on both sides before it is applied, and should be pressed on with a soft, dry cloth; then it will adhere so firmly that washing with soap and water will hardly remove it.

Sun-stroke.—This may be prevented by wearing a silk handkerchief in the crown of the hat.

Mortification.—To prevent wounds from mortifying, sprinkle sugar on them. The Turks wash fresh wounds with wine and sprinkle sugar on them. Obstinate ulcers may be cured with sugar dissolved in a strong decoction of walnut leaves.

An Excellent Remedy.—Saturate small pieces of rags of woolen material, (raveling of hose or flannel) with grease (lard or sweet oil), which place upon ignited wood, coal, or charcoal, in an iron kettle, so that they smoke without blazing. Hold the wound over the smoke—if convenient, covering the whole with a blanket, to condense the smoke upon the wound. The kettle should be in or near the chimney, or the windows open at the top, to prevent the deadly effect of inhaling the smoke. The same treatment is also good to prevent lockjaw.

For Cuts, Wounds, etc.—Take white pine turpentine, beeswax and Castile soap, of each one ounce, of bayberry tallow and mutton tallow each two ounces—(if the turpentine cannot be had, take one ounce of rosin instead)—and as much lard or fresh butter as will be sufficient to make it into a soft salve. Simmer all together in an earthen vessel till well dissolved. Do not burn it. By adding a little of the green bark of sweet elder and a little sweet oil it makes a good salve for a burn.

The buds of the balm of Gilead or elder flowers, infused in liquor, make it a good wash for cuts or wounds, to prevent soreness or inflammation.

For a Fish Bone in the Throat.—If any person should become choked with a fish bone, and cannot cough it out, or otherwise easily remove it from the throat, let him take the white of three or four eggs, and if that does not carry the bone down into the stomach (and perhaps it would be best that it should not carry the bone down), let him take immediately from a teaspoonful to a tablespoonful of ground mustard in a little warm water, which will be pretty certain to make him vomit almost immediately, and as the eggs are thrown from the stomach they will be very likely to carry the bone from the throat. If no mustard is on hand, take some other quick emetic.

For Scald Head. — This loathsome and painful disease will yield to the magical powers of the mustang liniment in an incredibly short time. The hair should be cut short and the scalp thoroughly washed with Castile soap and water, and then anoint with the liniment. A few applications will effect a cure.

Scurf in the Head. — *A Simple and Effectual Remedy.* — Into a pint of water drop a lump of fresh quick lime about the size of a walnut; let it stand all night, then pour the water off clear from the sediment or deposit, add a quarter of a pint of the best vinegar, and wash the head with the mixture. It is perfectly harmless. Only wet the roots of the hair.

For Salivation. — Take two parts of flour of sulphur and one part of cream of tartar; mix them into a kind of paste with a little honey or molasses, take a teaspoonful two or three times a day, or sufficient to operate slightly on the bowels, and use flour of sulphur freely in the mouth and on the gums.

Cure for Stammering. — At every syllable pronounced tap at the same time with one of the fingers, and by doing so the stammerer will soon be able to speak quite fluently.

For Indigestion. — Take one pint of rose water, six drachms of sulphate of magnesia, and one ounce of tincture of cascarilla; mix. Dose—three tablespoonsful twice a day.

Night Sweats. — Sage tea or sweet fern tea, drank cold on going to bed, is good to check night sweats.

For Gonorrhœa. — It is said that three small pills of gum turpentine every day, with eating clarified rosin throughout the day, will cure this disease.

A Good Tonic.—A gill of mustard seed and a handful of horse-radish root, put in a pint of wine, make a good tonic.

Teas.— There are two kinds of teas—the green and the black. Black tea is the more harmless, while green tea is the more exhilerating. In dyspepsia and nervous diseases the black tea is the better to use. It is a gentle astringent and carminative, allays sickness at the stomach, and produces composure of mind. Green tea, on the contrary, produces wakefulness, tremor, and nervous affections.

Seidlizt Powders.—The proper way to make these powders is—take two drachms of Rochelle salts, and two scruples of carbonate of soda (this is generally put into a blue paper), add half a drachm of tartaric acid (which is generally put into a white paper, as sold in the stores), mix them together in a large glass with water, and drink while in a state of effervescence. The quantity of Rochelle salts may be less than two drachms if they are not wanted to operate much on the bowels.

Godfrey's Cordial. —Dissolve two and a half drachms of sal tartar in three and a quarter pints of water, to which add one pint of thick sugar-house molasses, and afterwards three ounces of laudanum. Dissolve half a drachm of oil of sassafras in four ounces of alcohol, and add to the above. Shake well and it is ready for use.

Flaxseed Syrup.—This excellent remedy for a cough is made thus: boil one ounce of flaxseed in a quart of water for half an hour; strain and add to the liquid the juice of two lemons and half a pound of rock candy. If the cough is accompanied by weakness and a loss of appetite, add half an ounce of powdered gum arabic. Set this to simmer for half an hour, stirring it occasionally. Take a wine glass of it when the cough is troublesome.

PHYSIC FOR CHILDREN.—*Rhubarb and Magnesia.*—Mix one drachm of pulverized rhubarb with two drachms of carbonate of magnesia, and half a drachm of ginger. Dose—from fifteen grains to a drachm. Use as a purgative for children.

Another.—*Compound Soda.*—Mix one drachm of calomel, five drachms of sesqui-carbonate of soda, and ten drachms of compound chalk; pulverize together. Dose—five grains. Use as a mild purgative for children during teething.

THOMPSON'S HOT DROPS, OR "NUMBER SIX."—Take four pounds of gum myrrh, one pound of bayberry bark, twelve ounces of balmony, half a pound of skull-cap, five ounces Cayenne pepper, and five gallons of good brandy. Shake once or twice a day for eight or ten days, when it is fit for use. This is a stimulant and tonic, and an excellent embrocation for rheumatism, sprains, etc. Dose—from one to two teaspoonsful in a little warm water.

THOMPSON'S COMPOSITION POWDER.—Take six pounds of bayberry bark, three pounds of ginger, six ounces of Cayenne pepper, and six ounces of cloves; pulverize them, mix thoroughly and sift. Dose—a teaspoonful in half a pint of boiling water. Cool and sweeten.

EFFICACY OF ONIONS.—A writer says: "We are often troubled with severe coughs—the results of colds of long standing, which may turn to consumption or to premature death. Hard coughs cause sleepless nights, by the constant irritation in the throat, and a strong effort to throw off the offensive matter from the lungs. The remedy I propose has been tried by me, and often recommended to others with good results, which is simply to take into the stomach, before retiring for the night, a piece of raw onion, after chewing it. This esculent, in an uncooked state, is very heating, and tends to collect the waters from the lungs and throat, causing immediate relief to the patient. Sliced onion, in a raw state, will collect poison from the air, and also from the human system when taken internally or externally applied to the arm-pits.

EFFECTS OF ROASTED COFFEE. — The "London Medical Gazette" gives an account of the numerous experiments to ascertain the deodorizing effects of roasted coffee. It finds this material the most powerful means known, not only for rendering animal and vegetable effluvia innocuous, but of actually destroying them. A room in which meat in an advanced degree of decomposition had been kept for some time, was instantly deprived of all smell on an open coffee roaster being carried through it containing a pound of coffee, newly roasted. In another room, exposed to the effluvia occasioned by the clearing out of a cess-pool, so that sulphurated hydrogen and ammonia in great quantities could be chemically detected, the stench was completely removed in half a minute by three ounces of roasted coffee."

BRANDIES, WINES AND GINS.—*Their Medical Properties and Uses.*—Blackberry, raspberry and cherry wines, and blackberry and cherry brandies are astringent, a tonic, a remedy for dysentery, flux, and diarrhœa, either in children or adults, and are applicable to all other cases in which vegetable astringents are required. All are gentle and healthy stimulants.

Strawberry, Native Grape, and Lemon Cordial Wines produce action in the living tissues, expel and allay fever, diminish animal temperature, are cooling and soothing to the human body, and gentle and healthy stimulants.

Ginger Wine is a grateful stimulant and carminative; is very beneficial in cases of dyspepsia, flatulent colic, and a feeble state of the alimentary canal, attendant upon atonic gout, etc.

Port, Sherry, Madeira, Malaga, Muscat, and Claret Wines.— Sherry contains very little acid, and is preferable when the stomach is delicate, or exhibits a tendency to dyspeptic acidity.

Madeira is the most generous of white wines; is particularly adapted to the purpose of resuscitating debilitated constitutions, and of sustaining the sinking energies of the system in old age. It is an improper wine, however, for gouty persons.

Port, for cases of pure debility, especially when attended with a loose state of the bowels, unaccompanied with inflammation, acts as a powerful tonic, as well as a stimulant, giving increased activity to all the functions, especially to digestion.

Claret is much less heating, and is very useful on account of its aparient and diuretic qualities.

Champagne is very applicable to the sinking stage of low fevers, and most useful in the debility of the aged.

Muscat and Malaga are excellent and gentle stimulants, particularly for ladies and delicate persons.

Gins are diuretic. In cases of painful or prolonged menstruation gin will be found very beneficial. A mixture may be made of it with hot water and loaf sugar.

AN ENGLISH CURE FOR DRUNKENNESS.—There is a famous prescription in use in England, for the cure of drunkenness, by which thousands are said to have been assisted in recovering themselves. The receipt came into notoriety through the effects of John Vine Hall, commander of the steamship Great Eastern. He had fallen into such habitual drunkenness that his most earnest efforts to reclaim himself proved unavailing. At length he sought the advice of an eminent physician, who gave him a prescription which he followed faithfully for seven months, and at the end of that time had lost all desire for liquor, although he had been for many years led captive by a most debasing appetite. The receipt, which he afterwards published, is as follows: Five grains of sulphate of iron, ten grains of magnesia, eleven drachms of peppermint water; use twice a day. This preparation acts as a tonic and stimulant, and so partially supplies the place of accustomed liquor, and prevents the absolute physical and moral prostration that follows a sudden breaking off the use of stimulating drinks.

The following is said to be a cure for the appetite or disease of drunkenness: Five grains of sulphate of iron, ten grains of magnesia, eleven drachms of peppermint water, and one drachm of spirit of nutmeg, twice a day.

DOCTOR STRICKLAND'S ADVICE TO MOTHERS. — There are no diseases more distressing or common than those known as female complaints, and none more neglected. They are always aggravated and prolonged, if not generally produced, by neglect. This is not only the case with the young, but also with middle-aged women, both married and single. A slight cold will often produce the most serious results, if not taken in time, when the monthly discharges or menses are obstructed, or profuse or painful. You must know there is a cause for it, and it is your duty to take some course immediately to remedy the ill, either by applying to a physician, or making use of the following preparations, which will generally have the desired effect in nine out of ten cases:

For Obstructed Menstruation. — Take thirty grains of sulphate of iron, thirty grains of potassa (subcarbonate), thirty grains of white sugar, one drachm of myrrh; make them into pills of three and a half grains, two to be taken three times a day, when there is no fever present.

Another Recipe for the same. — Take one drachm each of powdered myrrh and powdered rhubarb, one scruple of the extract of aloes, two drachms of the extract of chamomile; mix with syrup, divide into moderate-sized pills, and take two or three a day.

To Promote the Menstrual Secretion. — Take pills of aloes and myrrh one drachm, compound iron pills sixty grains; mix, and form into twenty-five pills. Dose—one twice a day.

Another for the same.—Take compound galbanum pill one drachm, Socotrine aloes one drachm. Mix, and take one twice a day.

How to Check an immoderate flow of the Menses.—Take infusion of roses eight ounces, laudanum twenty drops. Two tablespoonsful three times a day.

Another for the same.—Take tincture of ergot one ounce, liquor of ammonia three drachms; mix. Dose—one teaspoonful in a little water three times a day.

Another for the same.—Take twenty drops of tincture of iron in a little water, three or four times a day. The dose may be increased to thirty drops. This is the most simple, and generally has the desired effect.

To Cure Painful Menstruation.—Take powdered rhubarb, powdered jalap, and powdered opium, of each one drachm; mix with syrup of popies, divide into one hundred pills, and take one night and morning.

How to Avoid Miasma.—The most favorable circumstances for the production of a miasmatic epidemic—speedy, malignant and wide-spreading—are the exposure of the muddy bottom of a pond or sluggish stream to the beaming heat of a summer sun. In less than a week whole neighborhoods have been stricken with disease, yet by the well established laws of miasm, five families may dwell within half a mile of a drained mill-pond, and only one will suffer from it, while the other four will remain exempt from unusual disease. First, if a rapid stream of considerable width runs between the drained pond and the house. Second, if there is interposed a thick hedge or growth of living, luxuriant trees or bushes. A treble row of sunflowers is known to have answered the purpose in repeated cases. Third, if the prevailing winds from June to October are from the house toward the pond. Fourth, if the house be on a steep hill.

The reasons for the above exemptions are here shortly

recapitulated: First, miasm does not cross a wide and rapid stream. Second, miasm is absorbed by thick, living, luxuriant foliage. Third, miasm cannot travel against the wind. Fourth, miasm cannot ascend a high, steep hill. There is no mystery in these variation, nor any complexity, when the laws of miasm are thoroughly understood.

It will be practically useful for the young farmer, in a pecuniary point of view, to understand, farther, that in one year a house on the banks of a mill-pond or sluggish stream may be visited.

Air in motion dissipates miasma. Hurricanes are therefore enemies to pestilence. Plagues often follow a long-continued quiet of the atmosphere.

STRYCHNINE.—This poison, that has lately become so notorious, is the production of the *strychnos nux vomica* tree, that grows in Ceylon, and in several districts in India. The tree is of moderate size, with thick, shining, green leaves, and a short, crooked stem. The tree bears orange-colored berries, about the size of a small apple, which have a hard, smooth rind, filled with a soft pulp. Many kinds of birds feed on this soft portion of the berry, as their favorite food, without any injury. But in the pulp are several flat seeds, covered with very minute silky hairs. This seed is the deadly poison nut. The bark of the tree is also poisonous. Here lies a danger that should be carefully watched. The deadly bark of the strychnos so nearly resembles in appearance and taste the Peruvian bark, that the former has been sold in Europe to a considerable extent, as the real quinine bark. Some fatal cases have resulted from this deception. American druggists should be very careful in ascertaining the quality of bark they sell as Peruvian.

THE UPAS TREE.—The story that the upas tree of the island of Java exhales a poisonous aroma, the breathing of which causes instant death, is now known to be false. The tree itself secretes a juice which is deadly poison, but its aroma or odor is harmless. Strychnine is made from the seeds of a species of the upas tree. The story that there is a poisoned valley in Japan where this tree grows is one. Such is the name of a district, the atmosphere of which produces death. The effect, however, is not occasioned by the upas tree, but by an extinct volcano near Batar, called Gueva Upas. From the old crater and the adjoining valley is exhaled carbonic gas, such as often extinguishes life in this country, in old wells and foul places. This deadly atmosphere kills every created thing which comes within its range—birds, beasts, and men. By a confusion of names, the poisonous effects of this deadly valley have been ascribed to the upas tree.

POISONS IN DAILY USE.—Ignorance very often conceals a deadly weapon in our choicest articles of food, but selfishness often conceals a greater. It manufactures and commends poisons for others in many temptingly disguised forms. Candies, toys and cakes are ornamented or colored with various poisons. Arsenite of copper and carbonate of copper are used in powder to ornament cake green or color candies. The blending in various ways, in candies and on cakes, makes them attractive to the eye, but destructive to the health of those who use them. Cakes ornamented with colored dust, candies colored in such nice style, and toys so highly attractive to children, cause decayed teeth, canker, intestinal inflammation, nauseating headache, colic, spasms, and oftentimes convulsions. Confectionery may be prepared without coloring material, so as to be wholesome. Gay colors are made of poisonous material, that ought never to be introduced into food or drinks. Wall paper, ornamented with beautiful green, pretty yellow and lively red, often diffuses, through sleeping and sitting rooms, an atmosphere impregnated with a poisonous vapor, that causes headache, nausea, dryness of the throat and mouth, cough, depression of spirits, prostration of strength, nervous affections, boils, watery swellings of the face, cutaneous affections, and inflammations of the eyes. These occur in more serious forms in apartments that are not constantly and thoroughly ventilated.

THE DUMB MADE TO SPEAK.—*Marvelous Effect of Whisky upon a Mute for a Quarter of a Century!*— The Cambridge City correspondent of the "Cincinnati Gazette" narrates the following curious incident:

"So seemingly miraculous has been the cure effected in the case of Miss Barnell, who was a mute for nearly twenty-six years, and whose speech was so singularly reproduced by the use of simple whisky, that the public appear loath to believe that it is anything but a hoax; and I am flooded with inquiries to know if there is any fact in the case, and if so, how the patient is doing, and if she continues to talk, etc.: therefore, I think a more minute diagnosis of the case might be of interest to your readers.

"It has now been two months since the speech of Miss B. was reproduced, and many of the knowing ones, both of the profession and of the people, predicted that as soon as the effects of the whisky would pass away, she would relapse into her former state. Yet such is not the case. She continues to talk as well as she has ever done since the return of her speech, with no visible impediment, and no external difficulty except an occasional movement of the hands, as formerly, when manipulating. Her general health has not entirely recovered from the unnatural shock produced by such a quantity of the stimuli — more than a pint having been used; but, aside from nervous prostration, which nature alone will overcome, she has as much vitality as usual.

"Singular as this case may appear to the public, it has been reported in an unvarnished form, as the hundreds who have visited her can verify, and the case would have an unparalleled renown had I been one of the profession. When the lady was seized with the malady she was only fourteen years of age, and could neither read, write, nor manipulate, and hence was entirely deprived of the ability to impart to her friends what transpired in her mind during the thirteen days she was in a trance. Thus she had first to manipulate, and then to read and write. The former she learned from some friends who had been educated at your State institution; the latter she has acquired so as to read with ease, and now that she can articulate, she feels that she is blest beyond measure. SAM. H. HOSHOUR."

OUR HAIR.—Doctor Dio Lewis has the following suggestions in relation to preserving the hair:

"God has covered the skull with hair. Some people shave it off. Mischievous practice! It exposes the brain. It exposes the throat and lungs—the eyes, likewise, say wise physiologists.

"Men become bald. Why? Because they wear close hats and caps. Women are never bald, except by disease. They do not wear close hats and caps. Men never lose a hair below where the hat touches the head, not until they have been bald for twenty years. The close hat holds the heat and perspiration: thereby the hair glands become weak and the hair falls out. What will restore it? Nothing after the scalp becomes shiny. But in process of falling out—recently lost—the following is best: Wash the head freely with cold water once or twice a day, and wear a thoroughly ventilated hat. This is the best means to arrest the loss and restore what is susceptible of restoration.

Cheap Wash for the Hair. — Dissolve over the fire, in a quart of water quite boiling, one ounce of borax and half an ounce of camphor, finely powdered. Allow it to cool, and perfume it, if you chose, with a little rose water. Wash the hair frequently with the preparation, and brush it until it dries.

Medical Use of Salt.—In many cases of disordered stomach a teaspoonful of salt is a certain cure. In the violent internal agony termed colic, add a teaspoonful of salt to a pint of cold water, drink it, and go to bed: it is one of the speediest remedies known. The same will revive a person who seems almost dead from a heavy fall.

In an apoplectic fit no time should be lost in pouring down water, if sufficient sensibility remains to allow of swallowing; if not, the head must be spunged with cold water till the sense returns, when salt will completely restore the patient from the lethargy.

In a fit the feet should be placed in warm water, with mustard added, and the legs briskly rubbed, all bandages removed from the neck, and a cool apartment procured, if possible. In many cases of severe bleeding at the lungs, and when other remedies fail, Doctor Rush found that two teaspoonsful of salt completely stayed the blood. In case of a bite from a mad dog, wash the part with a strong brine for an hour, and then bind on some salt with a rag.

In toothache, warm salt and water held to the part, and removed two or three times, will remove it in most cases. If the gums be affected, wash the mouth with brine. If the teeth be covered with tartar, wash them twice a day with salt and water.

In swelled neck, wash the part with brine, and drink it also twice a day until cured.

Salt will expel worms, if used in food in a moderate degree, and aid digestion; but salt meat is injurious, if used much.

Medical Flora.—*Alder.*—A wine glass of syrup made of a decoction of black alder and honey, and drank three or four times a day, is good for affections of the lungs, and tea or beer of the bark of the root of this alder is good to purify the blood.

Balm of Gilead.—The buds of this tree, infused in spirits, are good for coughs, and also for bathing wounds.

Blackberry.—This is an astringent, and a syrup made of the juice of the berries is an excellent remedy for diarrhœa, dysentery, and summer complaint.

Black Snake Root—(Rattle Weed—Black Cohosh.)—A decoction of the root makes a good wash for all kinds of inflammation, and a tincture of it is good for chronic rheumatism. A tea of it is excellent, both as a preventive and cure of small-pox.

Boneset.—This is a purgative, tonic, and sweating. This plant possesses various properties, according to the dose in which it is administered. In dyspepsia it is an excellent tonic, if the tea is taken cold and in small quantities through the day. A decoction of the leaves and flowers taken while *warm*, in small doses, and repeated frequently, produces free perspiration; and if taken in large doses, it will evacuate the stomach in a safe and gentle manner; and if administered cold, it acts as a tonic and laxative. It is also useful in coughs, colds, and pulmonary complaints.

Burdock.—This root, in a tea or syrup, is an excellent purifier of the blood.

Catnip.—In colds, a tea made of it, and drank while warm, generally produces a free perspiration. In fevers it promotes perspiration without increasing the heat of the body.

Comfry.—A syrup made of this plant is good for coughs and colds, and also for consumptive complaints.

Chamomile.—The flowers alone are used, either in tea or an infusion, and which, if drank cold, is a good tonic. The flowers boiled with milk, made into a poultice, and applied to the neck, are good to remove painful glandular swellings, which often arise from cold.

Centaury.—This plant makes a good bitter, which is valuable to strengthen the stomach.

Dandelion.—This is a tonic and diuretic, an excellent corrector of the bile, and a valuable remedy in hepatic diseases.

Dittany, Mountain.— A tea of this plant is good for colds, nervous headache, hysterics, and stoppage of the urine.

Dogwood.— A tea of the bark of this tree is an excellent tonic. The bark of this tree and the bark of wild cherry tree, in equal parts, boiled together, so as to make a strong tea, and sweetened, with a little French brandy added, makes a very good tonic bitter for weakly persons and for general debility.

Elecampane.—The root of this plant, when combined with spikenard root, burdock root, sarsaparilla root, and hoarhound tops, in equal parts, and boiled together, so as to form a strong tea, and sweetened with honey, makes an excellent syrup for all lung diseases, and is also good for colds and coughs.

Elder.— The flowers of the elder are good to purify the blood; and, infused in rectified spirits of wine, with a little gum camphor, make an excellent remedy for weak eyes. A little rubbed around the eyes and on the temples two or three times a day, is very strengthening for weak eyes. The leaves strewn among grain, when put into the bin, will protect it from the ravages of the weevil; and if placed in the branches of the plum and other fruit trees, or scattered over plants, will protect them from the ravages of insects. They also make an excellent ointment for eruptions of the skin.

Elm, Slippery.—The bark of this tree is mucilaginous and cooling, and good in inflammation of the stomach, lungs, and bowels; it also makes an excellent wash for sore mouth and sore throat. Infused in water it makes a good drink in diarrhœa and dysentery, and in all internal inflammation, where water alone may be drank. The bark, pulverized, makes an excellent poultice for all tumors, ulcers, sores, burns, swellings, wounds and inflammations.

Fennel Seed.— A tea made of this seed is good for colic in infants, and also to expel wind.

Garlic.—A syrup made from this is good for coughs and inflammation of the lungs. A tea of it is good for infants.

Ginger. — A tea of this, taken warm, is good to relieve pain caused by wind or cold.

Ginseng. — This root is tonic and nervine, and useful in cases of debility, loss of appetite and dyspepsia.

Ground Ivy. — A decoction or syrup made from the leaves of this plant is good for coughs, colds and consumption, and to purify the blood.

Golden Seal — (*Yellow Root* — *Yellow Pacoon*). — This is tonic and laxative, and valuable in dyspepsia, and for sore mouth and sore throat.

Hops. — Steeped in vinegar, and applied hot, are good to relieve pain and reduce swellings.

Horsh radish. — This root, pounded and applied, is said to be good for neuralgia. The leaves wilted and applied to the bowels relieve colic; and applied to cold feet, they produce warmth and circulation there; they are also good to reduce swellings.

Hyssop. — This makes a good tea for colds.

Hoarhound. — A strong tea made from this herb, sweetened with honey or molasses, is good for all colds, coughs, and pulmonary complaints.

Horsemint. — This is a powerful diuretic. A strong tea, aided by the warm bath, generally affords immediate relief in gravel and stoppage of the urine.

Life Everlasting. — A tea made from this plant is good for pain in the breast, weakness of the lungs, and other pulmonary diseases.

May-Apple — (*Mandrake*). — The root of this plant is an active cathartic. It is also good to put into old ulcers, to make them suppurate and discharge all their rotten and decayed matter.

Mullen. — A decoction of the leaves, used hot, makes a good fomentation to reduce swellings; and mixed with lard, makes a good ointment for piles.

Parsley. — A tea made of this is good for stoppage of the water.

Plantain. — The leaves simmered in fresh butter make a good ointment for tetter and salt rheum. The juice of the leaves is said to be good for bites of snakes and insects.

Prickly Ash. — The bark and berries of this tree are stimulant and tonic, and are a valuable remedy for ague and fever, dyspepsia, cold hands and feet, drowsiness, and all affections dependent on a sluggish circulation.

Pennyroyal. — This is stimulant, and a tea made of it, and drank warm, promotes perspiration, and is good for colds.

Peppermint. — This is a stimulant, and a tea of it drank warm promotes perspiration, and may be given in all cases of colds, pain in the stomach and bowels, headache and dyspepsia.

Peruvian Bark. — This bark, infused in good liquor, makes an excellent tonic bitter, and is good for chills and fever, and general debility.

Rattle Weed. — See "Black Snake Root."

Sage. — A tea made of this herb, and drank warm, is an excellent remedy to promote perspiration, and is valuable in colds, coughs and fevers.

Sassafras. — The bark of the root of sassafras is stimulant, and a tea of it is an excellent purifier of the blood, and is valuable in rheumatism.

Sumach. — The leaves and berries are stimulant and astringent, are used in strangury and dysentery, and for making a wash for old sores.

Slippery Elm. — See "Elm, Slippery."

Spearmint. — This is stimulant and tonic, and a tea of it promotes perspiration, relieves sickness at the stomach and pain in the stomach and bowels, and is an excellent remedy for the gravel and stoppage of the urine.

Smart Weed. — A tea of this herb is good to stop vomiting; when drank freely will produce profuse sweating, and is good to break up a cold when threatened with fever. This weed simmered in vinegar makes an excellent decoction to reduce swellings, by applying it as a fomentation.

Senna.—This is purgative, and is usually combined with manna; and when a small portion of coriander seed is added, to prevent its griping effects, it is a mild and useful cathartic.

Spikenard.—A syrup made from the root of this plant, and sweetened with honey, is good for colds, coughs, and pulmonary diseases.

Sweet Fern.—A tea made of this, and drank freely, is an excellent remedy in dysentery or bloody flux. The tea drank cold is good to check night sweats

Sarsaparilla.—This root makes an excellent syrup or decoction for purifying the blood.

Seneca Snake Root.—A decoction of this root is good for hives, croup, and dropsy.

Tansy.—This herb makes a good tonic bitter, and if taken in the form of warm tea it will produce sweating.

Virginia Snake Root.—A tea made from this root, and drank freely while warm, produces perspiration; if taken cold, and in small quantities, it is a good tonic.

Water Plantain.—The roots of this plant boiled until soft, made into a poultice, and applied, are good to remove inflammations, reduce swellings, and cleanse and heal inveterate ulcers.

Wild Cherry.—The bark of this tree is tonic, and when combined with poplar root bark and dogwood bark, forms a good strengthening bitter. They should always be boiled together, and when cold, good spirits should be added to the decoction.

A syrup made from the bark of this tree is very good for coughs, colds, and pulmonary diseases.

Winter Green—(Pipsissewa).—This is stimulant and tonic, and useful in scrofula and diseases of the kidneys. This herb or plant infused in good rye whisky is an excellent remedy for the cure of rheumatism, when taken two or three times a day.

Yellow Root.—See "Golden Seal."

MISCELLANEOUS DEPARTMENT.

MISCELLANEOUS.

BAROMETERS. — *Leech Barometer.* — Take an eight-ounce phial, put into it three gills of water, and place therein a healthy leech, changing the water in summer once a week, and in winter once a fortnight, and it will most accurately prognosticate the weather. If the weather is to be fine, the leech lies motionless at the bottom of the glass, and coiled together in a spiral form; if rain may be expected, it will creep up to the top of its lodgings, and remain there till the weather is settled; if we are to have wind, it will move through its habitation with amazing swiftness, and seldom goes to rest until it begins to blow hard; if a remarkable storm of thunder and rain is to succeed, it will lodge for some days previous, almost continually out of the water, and will discover great uneasiness in violent throes and convulsive motions; in frosty, as in clear, summer-like weather, it lies constantly at the bottom; and in snow, as in rainy weather, it pitches its dwelling in the very mouth of the phial. The top should be covered over with a piece of muslin.

The Farmer's Barometer. — Take a common glass pickle bottle, wide-mouthed, fill it within three inches of the top with water; then take a common Florence oil flask, remove the straw covering, cleanse the flask thoroughly, place the flask into the bottle as far as it will go, and the barometer is complete. In fine weather the water will rise into the neck of the flask, even higher than the mouth of the pickle bottle; and in wet, windy weather it will fall to within an inch of the mouth of the flask. Before a heavy gale of wind the water has been seen to leave the flask altogether, at least eight hours before the gale comes to its height.

A Truthful and Cheap Barometer.—Take a clean glass bottle and put into it a small quantity of pulverized alum, then fill up the bottle with spirits of wine. The alum will be perfectly dissolved by the alcohol, and in clear weather the liquid will be as transparent as the purest water. On the approach of rain, or in cloudy weather, the alum will be visible in a flaky, spiral cloud in the centre of the fluid, reaching from the bottom to the surface. This is a cheap, simple, and beautiful barometer, and is placed within the reach of all who wish to possess one. "For simplicity of construction," the "Scientific American" says, "this is altogether superior to the frog barometer, in general use in Germany."

WHITEWASH. — *Very Nice for Rooms.*—Take four pounds of whiting, and four ounces of white or common glue. Let the glue stand in cold water over night. Mix the whiting with cold water, heat the glue until dissolved, and pour it into the other while hot. Make a proper consistency, then apply.

Whitewash that will not Rub Off. — Mix half a pail of lime and water, ready for whitewashing; take half a pint of flour and make a starch of it, and pour it into the whitewash while hot; stir it well, and it is ready for use. This will not rub off.

Brilliant Whitewash. — Such as is used on the east end of the President's house at Washington is made thus: Take half a bushel of nice unslaked lime, slake it with boiling water, cover it during the process, to keep in the steam, strain the liquid through a fine sieve or strainer, and add to it a peck of salt, previously well dissolved in warm water, three pounds of ground rice, boiled to a thin paste, and stirred in boiling hot, half a pound of powdered Spanish whiting, and a pound of clean glue which has been previously dissolved by soaking it well, and then hanging it over a slow fire, in a small kettle placed inside a large one filled with water; add five gallons of hot water to the mixture, stir it well and let it stand a few days, covered from the dirt. It should be put on hot. It is said that about a pint of this mixture will cover a square yard on the outside of a house, if properly applied.

To Make a Brilliant Whitewash for all Buildings, inside and out.—Take clean lumps of well-burnt lime, slaked; add one fourth of a pound of whiting or burnt alum, pulverized, one pound of loaf sugar, three quarts of rye flour, made into a thin and well-boiled paste, and one pound of the cleanest glue, dissolved as cabinet-makers do. This may be put on cold within doors, but not outside. It will be as brilliant as plaster of Paris, and retain the brightness for many years.

Whitewash for Fences and Out-houses.—Half a bushel of unslaked lime, one peck of salt, three pounds of ground rice, half a pound of powdered whiting, one pound of dissolved glue; slake the lime in boiling water, in a covered vessel, strain through a wire sieve, add the salt, dissolved in hot water, and while hot the rice boiled to a thin paste, then the glue and whiting; let it stand for several days, and put it on hot.

PAINT.—*Fire and Water Proof.*—Take a sufficient quantity of water for use, and add as much potash as can be dissolved therein; when the water will dissolve no more potash, stir into the solution first a quantity of flour paste of the consistency of painters' size, then enough of pure clay to make it the consistency of cream. Apply with a painters' brush.

To Clean Paint that is not Varnished.—Take a flannel, and squeeze nearly dry, out of warm water, and dip in a little whiting; apply to the paint, and with a little rubbing, will remove grease, smoke or other soil. Wash with warm water, and rub dry with a soft cloth. It will not injure the most delicate color, and makes it look as well as new; besides it preserves the paint much longer than if cleaned with soap and water.

CEMENT.—*An Excellent Cement.*—The following is from the "Scientific American:" Five years ago, we applied a cement, composed of white-lead paint, whiting, and dry white sand, to a small tin roof that leaked like a sieve: it soon became nearly as hard as stone, has never scaled off, and has kept the roof since then perfectly tight. It was put on about the consistency of thin putty. Slater's cement for stopping leaks around chimneys is composed of linseed-oil, whiting, ground glass, and some brick-dust. It is a good cement for this purpose; also for closing the joints of stone steps to houses.

Hard Cement.—The following cement has been used with much success in covering terraces, lining cisterns and uniting stone flagging: Take 90 parts by weight of well-burned brick reduced to powder, add 7 parts of litharge, mix them together, and render them plastic with linseed oil. It is then applied in the manner of plaster; the body that is to be covered being again previously wetted on the outside with a sponge. When the cement is extended over a large surface, it sometimes dries with flaws in it, which must be filled up with a fresh quantity. In three or four days it becomes firm.

To make a Fire and Water Proof Cement.—Put together milk and vinegar, each half a pint. Separate the curd, and mix the whey with the whites of five eggs, beat well together, and sift into it enough unslaked lime to make it the consistency of thick paste. Broken vessels mended with this cement will not separate, for it resists both fire and water.

Japanese Cement.—This elegant cement is made by mixing rice flour intimately with cold water, and then gently boiling it. It is beautifully white, and dries almost transparent.

Cheap Cement.—Quick-lime, one part; white of egg, two parts. Mix, and apply at once.

Cracks in Stoves.—To close cracks in stoves through which air or smoke penetrates, apply while the stove is hot or cold, a mixture of common salt and fresh wood-ashes made into a paste with water.

Cement for Broken Glass, etc.—A bit of isinglass, dissolved in gin, or boiled in spirits of wine, is said to make strong cement for broken glass, china, and sea-shells.

LIQUID SOLDER.—*For Soldering and Cleaning Tin.*—Put into an open mouth vial one or two ounces muriatic acid. Then add zinc as long as it keeps boiling.

Directions for using.—Drop two or three drops where you wish to solder, then put a piece of soft solder on the place, then hold over a lamp or coals till melted.

RUST.—*To Remove Rust from Polished Iron.*—The best method for removing rust from a polished grate, is to scrape down to a fine powder some bath-brick; put it into a little oil, and rub the spots well with a piece of flannel dipped in the mixture, after which apply some whiting, also well rubbed in. This process must be repeated daily until all trace of the rust has disappeared. To prevent the grate or fire-irons from becoming spotted with rust, it is a good plan to rub them over with the fat from the inside of a fowl, and finish them off with whiting.

To Remove Rust from Knives, etc.—Cover the knives, etc., with sweet oil, and rub it on well; after two days take a lump of fresh quick lime, and rub till all the rust disappears.

The oil and lime form a sort of soap, which carries off all the rust. If new steel articles are rubbed well with oil, and not polished off until twenty-four hours after, they do not rust nearly so soon.

Mixture for Polishing Brass.—Spirits of turpentine, half a pint; rotten-stone, quarter pound; charcoal in powder, quarter pound. Mix well, and add quarter of a pint of sweet oil; finish with dry charcoal dust.

If your flat-irons are rough and smoky, lay a little fine salt on a flat surface, and rub them well; it will prevent them from sticking to any thing starched, and make them smooth.

To Prevent Metals from Rusting.—Melt together three parts of lard and one of rosin, and apply a very thin coating. It will preserve Russia iron stoves and grates from rusting during summer, even in damp situations.

To Take Rust out of Steel.—Rub well with sweet oil, and let the oil remain upon them for forty-eight hours. Then rub with leather sprinkled with unslacked lime, finely powdered, until the rust disappears.

To Clean Freestone.—Wash the hearth with soap, and wipe it with a wet cloth. Or rub it over with a little freestone powder, after washing the hearth in hot water. Brush off the powder when dry.

VERMIN.—*Destruction to House Bugs.*—The French Academy of Science is assured, by Baron Thenard, that boiling soap and water, consisting of 2 parts of common soap, and 100 parts of water by weight, infallibly destroys bugs and their eggs. It is enough to wash walls, wood-work, etc., with the boiling solution, to be entirely relieved from this horrid pest.

Camphor a Remedy for Mice.—Any one desirous of keeping seeds from the depradations of mice, can do so by mixing pieces of camphor gum in with the seeds. Camphor placed in drawers or trunks will prevent mice from doing them injury. This little animal objects to the odor, and keeps a good distance from it.

A writer in the "Scientific American" says: "A good plan to destroy roaches, without the danger of using poison, is, to fill a basin or a similar vessel about two-thirds full of water sweetened with molasses, and set in a corner where they most frequent at night, and where they can get on the vessel; you will find in the morning as many as the liquid will drown. I have rid my house of them in this way, destroying hundreds in a night.

Salt for Bed Bugs.—An exchange says: "By washing bedsteads with salt water, and filling the cracks where bed bugs frequent, with salt, those troublesome vermin can be speedily and cheaply got rid of. The salt is said to be inimical to them and they will not trail through it. It is certainly worth a trial."

To Destroy Red Ants.—Place a dish of cracked shagbarks (of which they are more fond than of anything else); they will gather upon it in troops; put some corrosive sublimate in a cup; take the dish containing the shagbarks and ants and throw them into the fire, and with a feather, sweep those that may be left, into the cup, and wet all the cracks from whence they came with the corrosive sublimate. When this has been repeated four or five times the house will be effectually cleared.

To get rid of Bed and other Bugs.—Gather a handful of smartweed, boil in a pint of water, and when cold rub the liquid where they frequent, and they will soon disappear.

To preserve Houses from Vermin.—Put half a drachm of corrosive sublimate, with a quarter of an ounce of spirits of salts, into one quart of spirits of turpentine. Shake well; with this wash the places where bugs resort—a sure exterminator. It is an active poison.

Cedar Chests are best to keep flannels, for cloth moths are never found in them. Red cedar chips and gum camphor are good to keep in drawers, wardrobes, closets, trunks, etc., to keep out moths.

In laying up furs for summer, lay a tallow candle in or near them, and danger from worms will be obviated.

To Remove Flies from Rooms.—Take half a teaspoonful of black pepper in powder, one teaspoonful of brown sugar, and one tablespoonful of cream—mix them well together, and place them in the room, on a plate, where the flies are troublesome. They will soon disappear.

Tomato Worms.—The Port Byron "Times" says that several persons near Auburn have recently been fatally stung by a large worm that infests tomato vines, death ensuing within a few hours. A lady in Port Byron discovered one of these monsters on her tomato vines and narrowly escaped being stung. The worm is described as about three inches long, of a green color, and armed with claws and nippers, with a black horn extending in front some three-fourths of an inch long.

For Preserving Leather.—*Liquid Blacking.*— Mix a quarter of a pound of ivory-black, six gills of vinegar, a tablespoonful of sweet oil, and two of molasses. Stir the whole well together, and it is fit for use.

Oil Paste Blacking.—Take oil of vitriol two ounces, tanners' oil five ounces, ivory black one pound, molasses five ounces; mix the oil and vitriol together and let it stand a day, and then add the ivory black and molasses and the white of an egg, and stir it well together to a thick paste. This is a superior blacking, will not injure the leather, and gives universal satisfaction.

Water Proof for Leather.— Take linseed oil one pint, yellow wax and white turpentine, each two ounces, Burgundy pitch one ounce, melt and color with lamp-black.

To make Boots water proof.—Yellow beeswax, Burgundy pitch, and turpentine, of each two ounces; boiled linseed oil one pint. Apply to the boots with the hand before the fire, till well saturated.

To render Shoes water proof.—Warm a little beeswax and mutton suet until it is liquid, and rub some of it slightly over the edges of the sole where the stitches are.

Water proof Composition for Leather.—One-half pound tallow, two ounces turpentine, two ounces beeswax, two ounces olive oil, four ounces hog's lard. Melt the materials by a gentle heat. Rub the mixture on the leather a few hours before using. It should be rubbed on new boots or shoes two or three times before using them. By adding a small quantity of lamp-black and increasing the quantity of beeswax, an excellent *black-ball* is obtained.

INKS.—*Blue Ink.*—Powdered Prussian blue, 1 oz.; concentrated muriatic acid, 1½ to 2 oz.; mix in a glass bottle, and after 24 to 30 hours, dilute the mass with a sufficient quantity of water. The Prussian blue must be a pure article.

To make durable Black Ink.—Take four pounds of nut galls, powdered, one pound of gum arabic, fourteen ounces of copperas, and one gallon of soft water; mix well together. This ink will endure for centuries.

Durable Ink for Marking Linen.—Dissolve a couple of drachms of lunar caustic, and half an ounce of gum arabic in a gill of rain water. Wet the articles in the place where they are to be marked with strong saleratus water; let them get perfectly dry, then iron them smooth. The heat of the iron will turn them a dark yellow, as if scorched. Washing will efface it, but it should not be done till after the marking is thoroughly dry; place it in the sun or near a fire to dry, after marking. The ink should be kept in a small vial; the water will evaporate when it is kept long. In such a case add more water, and shake it up well in the bottle when you wish to use it.

The Hair—*How to color it Black.*—An English writer states that a liquid that will color the human hair black, and not stain the skin, may be made by taking one part of bay rum, three parts of olive oil, and one part of good brandy, dry measure. The hair must be washed with this mixture every morning, and in a short time the use of it will make the hair a beautiful black, without injuring it in the least. The article must be of the best quality, mixed in a bottle, and always shaken well before being applied.

Another Method.—Take vinegar, lemon-juice, and powdered litharge, in equal parts; boil slowly for half an hour; wet the hair with this decoction, and in a short time it will turn black.

To destroy superfluous Hair.—Take fresh stone lime, one ounce; pure potash, one drachm; sulphuret, one drachm. Reduce them to a fine powder in an earthen or glass mortar, and add enough soft water to make a thin paste. Then wash the hair in warm water, and apply the paste, by rubbing gently a little on the spot where you wish to remove the hair. As soon as the skin is much reddened, wash it off with strong vinegar. Do not let it remain on more than three to five minutes. Wash the place with a flannel cloth, and the hair will be removed. The skin will be softened and improved in appearance.

To Make Wood Fire or Water Proof.—Take some gravelly earth, wash it clean from all heterogeneous matter, and dissolve it in a strong solution of caustic alkali. Spread this on the wood.

To Make Paper or Cloth Fire Proof.—Dip it in a strong solution of alum water, and then dry it thoroughly. Neither the color nor quality of the paper will be affected. In this it will be fire proof.

To Renovate Manuscripts.—Take a hair pencil and wash the part which has been effaced with a solution of prussiate of potash in water, and the writing will again appear if the paper has not been destroyed.

To Sweeten Musty Casks.—Fill the cask with boiling water, and then put in some pieces of unslaked lime, keeping up the ebullition half an hour. Then bung it down, and keep it until cold, when turn it out, and rinse well with water.

To Prevent the Smoking of a Lamp.—Soak the wick in strong vinegar, and well dry it before you use it.

How to Preserve a Bouquet.—When you receive a bouquet, sprinkle it lightly with fresh water; then put it in a vessel containing some soap-suds, which nourish the roots, and keep the flowers as good as new. Take the bouquet out of the suds every morning, and lay it sideways in fresh water, the stock entering first into the water; keep it there a minute or two, then take it out, and sprinkle the flowers lightly by the hand with pure water. Replace the bouquet in the soap-suds, and the flowers bloom up as fresh as when gathered.

Cisterns and Wells.—It is probably not known by most persons that the bottom of a cistern, or deep well, even, may be thoroughly inspected for filth or lost articles by using a common mirror (looking-glass). When the sun shines, hold the mirror so that the light will be reflected in a bright spot at the bottom of the water, and a pin can be seen at a depth of from ten to twenty feet, or more. We have in this manner seen fishes at the bottom of from thirty to forty feet of clear water. If the sun be hid by intervening objects, use two or more mirrors to bend the light by double or triple to the desired point. We have thus thrown the light, coming into the dining-room window, by one mirror, through the door into the kitchen, by another, to a corner of the latter room, and by a third mirror, cast it down into a cistern, sufficiently strong to see a small angle-worm.

LIQUID GLUE.—Break the glue in small pieces, then add vinegar—say two-thirds vinegar and one-third glue—shake it well several times during twenty-four hours, and it is fit for use.

HARD PUTTY, around broken window panes, is quickly softened by pouring kerosene oil on it.

CURIOUS MODE OF SILVERING IVORY.—Immerse the ivory in a weak solution of nitrate of silver, and let it remain till it has acquired a deep yellow color. Then immerse it in clean water for a few moments, and expose it to the rays of the sun. In about three hours the ivory becomes black; but the black surface, on being rubbed, is at once changed to a brilliant silver.

TO MAKE LIME-WATER.—Take two tablespoonsful of unslaked lime, and put to it three quarts of boiling water, which will give two quarts of clear lime-water. Should any person wish to make a quantity of the lime-water, they can do so, taking of lime and water the proportions as directed above, keeping it in a stone jar, ready for use.

BREATH TAINTED BY ONIONS.—Leaves of parsley, eaten with vinegar, will prevent the disagreeable consequences of eating onions.

TO EXTINGUISH A CHIMNEY ON FIRE.—Put into the fire in the grate, or fire-place, or in the stove, a quantity of sulphur. Continue to burn sulphur until the fire in the chimney is put out. It will require only a few moments.

COMPOSITION FOR MAKING COLORED DRAWINGS AND PAINTS

RESEMBLE PAINTINGS IN OIL.—Take of Canada balsam, one ounce; spirits of turpentine, two ounces: mix them together. The drawing or paint should be first sized with a solution of isinglass in water, and when dry, apply the above with a camel's-hair brush.

TO BREAK GLASS ANY REQUIRED WAY.—Dip a piece of worsted thread in spirits of turpentine, wrap it round the glass in the direction required to be broken: then set fire to the thread.

WHEN cloths have acquired an unpleasant odor by being excluded from the air, charcoal, laid in the folds, will soon remove it.

TO DESTROY FOUL SMELLS—One pound of green copperas, costing only seven cents, dissolved in about one quart of water and poured down a privy, will effectually concentrate and destroy the foulest smells. For water-closets aboard ships and steamboats, about hotels, and other public places, there is nothing so nice and simple for cleansing as green copperas, dissolved under the bed, in any thing that will hold water, and thus render a hospital, or other places for the sick, free from unpleasant smells. For butchers' stalls, fish-markets, slaughter-houses, sinks, and wherever there are offensive putrid gases, dissolve copperas and sprinkle it about; and in a few days the smell will pass away. If a cat, rat, or mouse dies about the house, and sends forth an offensive gas, place some dissolved copperas in an open vessel near the place where the nuisance is and it will soon purify the atmosphere.

Chloride of lime is better to scatter about in damp yards and cellars, to purify them.

CLEANSING CLOTH, ETC.—*To extract Grease Spots from Silks and Colored Muslins.*—Scrape French chalk, put it on the grease spot, and hold it near the fire, or over a warm iron, or water-plate, filled with boiling water. The grease will melt, and the French chalk absorb it; brush or rub it off. Repeat if necessary.

To remove Mildew from Linen—Moisten a piece of soap, and rub it thickly into the part affected; then scrape fine whiting, and rub that in also. Lay the linen on the grass, and from time to time, as it becomes dry, wet it a little. If the spots are not quite removed, repeat the process.

To remove Ink Spots.—Wet the place immediately with sorrel or lemon juice, and rub on it hard white soap. Ink or iron mould may be removed by holding over a vessel of boiling water, and squeezing on the spot juice of sorrel, then rubbing with dry salt.

To remove Grease from Woolen or Silk.—Clay is never injurious. It should be moistened with boiling water, and when cold laid on the spot damp; it will draw out oil, and when brushed off, leave the garment uninjured.

To remove Spots of Grease or Paint from Woolen Garments. —Wet the spot with a few drops of benzine and rub it quickly between the fingers. Oil spots and stains from candle snuffs, on woolen table covers, paint spots on garments, etc., are thus perfectly removed without the slightest discoloration.

Ink Stains.—Recent stains of ink may be removed if before the ink is dry the places be washed with sweet milk; if this does not succeed, rub the spots with vinegar, lemon juice or tartaric acid, and afterward wash it with soap and water. Ink or iron stains may also be removed by the bleaching liquid already described.

Removing Stains.—The "Country Gentleman" says: "All clothes subject to be stained, such as table linen, napkins, children's clothes, towels, etc., ought to be examined before being put into any wash mixture or soap suds, as these render the stain permanent. Many stains will yield to good washing in pure soft warm water. Alcohol will remove almost any discoloration. Almost any stain or iron mold, or mildew, may be removed by dipping in a moderately strong solution of citric acid, then covered with salt and kept in the sun. This may require to be repeated many times, but with us has never failed."

To remove Stains from Silk.—Salts of ammonia mixed with lime will take out the stains of wine from silk. Spirits of turpentine, alcohol, and clear ammonia are all good to remove the stains from colored silks.

To remove Ink from Carpets, Furniture, etc.—Wipe the spot with oxalic acid; let it remain a few minutes, then rub it with a cloth wet with warm water. Colored paint, mahogany, and carpets will require washing with hartshorn water to restire the original color.

Removing Indelible Ink stains.—To remove spots of nitrate of silver indelible ink, moisten them for a few moments with moist chloride of lime, which forms chloride of silver, and then dissolve the latter by caustic ammonia. It may be sometimes necessary to repeat the operation. Cyanide of potassium may also be employed.

To remove Grease from Silk.—Wash the spots with ether.

Ink Stains should never be put into soapy or soda-water, or ley, as they directly become iron molds; but should be instantly wetted with clean water, and may be at once removed by the application of a little lemon juice or salt of lemon.

To remove Ink Stains.—Dip the spotted part into hot tallow. Then wash and the spots will disappear.

To restore the color to Silk.—Silks that have changed their color by acids can be restored by the free use of hartshorn.

To remove stains from Broadcloth.—Take an ounce of pipeclay, ground fine, mix it with twelve drops of alcohol and the same quantity of spirits of turpentine. Whenever it is needed, moisten a small portion of this mixture with alcohol; rub it on the spots and let it remain until dry; then rub it off with a woolen cloth, and the spots will disappear.

To take Ink spots out of Linen.—As soon as the accident happens, wet the place with the juice of sorrel or lemon, or with sharp vinegar, and the best hard white soap.

To extract Paint from Garments.—Saturate the spots with spirits of turpentine, let it remain a number of hours, and then rub it between the hands; it will crumble away without injury either to the texture or color of any kind of woolen, cotton or silk goods.

Good preparation for cleaning cloth.—Oil of turpentine, two ounces; borax, one ounce; strong liquid ammonia, three ounces. Mix, and use with a sponge.

To cleanse Gloves without Wetting.—Lay the gloves upon a clean board, make a mixture of dried fulling earth and powdered alum, and pass them over on each side with a common stiff brush; then sweep it off, and sprinkle them well with dry bran and whiting, and dust them well; this, if they be not exceedingly greasy, will render them quite clean; but if they are much soiled, take out the grease with crumbs of toasted bread, and powder of burnt bone; then pass over them with a woolen cloth dipped in fulling earth or alum powder, and in this manner they can be cleansed without wetting, which frequently shrinks and spoils them.

To clean Kid Gloves.—Light kid gloves can be cleaned by rubbing on cream of tartar, magnesia, or camphene, with a flannel cloth. Windsor soap and milk are also good.

DYES.—*To dye Nankeen color.*—A pailful of ley, with a piece of copperas half as big as a hen's egg boiled in it will color a fine nankeen color, which will never wash out. This is very useful for the linings of bed quilts, comforters, etc.

To dye Black.—Rusty nails or any rusty iron boiled in vinegar, with a small bit of copperas, makes a good black.

To dye Purple.—Cut a pumpkin so as to form a lid, take

out the inside and fill it with white yarn hanks or wool and pokeberry juice, set in a warm place till fermentation takes place, wash out in soap, and you have a beautiful royal purple, indelible. The fermentation sets the dyes, and it will take place in 8 or 10 days by the kitchen fire.

STAINS.—*To remove Stains.*—To remove acid stains from linen or cotton goods, moisten the cloth with water and hold a lighted match under the stain. The sulphurous gas from the match will remove the stain.

MUCILAGE.—*To make Mucilage.*—Put one ounce of best gum arabic into two ounces of soft water, and when it is dissolved it is ready for use.

To separate Papers stuck together with Mucilage, etc.—To separate papers that are stuck together with glue, mucilage, etc., as a scrap-book, etc., dip the leaves into hot water, and before they become dry, you can readily separate the pieces.

FOUL AIR.—*To remove Foul Air from Wells.*—A quantity of burnt but unslaked lime, thrown down into the water in the well, sets free a great amount of heat in the water and lime, which, rushing upwards, carries the deleterious gases with it. Then light a candle and let it down to the water or bottom of the well, and if it continues to burn freely descent can be made with safety, but if it goes out in the well, more lime must be thrown down before the descent is made.

BROWNING.—*To brown Gun Barrels and other Metals.*—Take tincture of iodine and dilute it with half its amount of water, and apply it with a clean rag; let it remain five or six hours, then brush the metal and rub it over with beeswax dissolved in turpentine, and it is done.

THE ART OF SWIMMING.—Men are drowned by raising their arms above water, the unbuoyed weight of which depresses the head. Other animals have neither motion nor ability to act in a similar manner, and therefore swim naturally. When a man falls into deep water he will rise to the surface, and will continue there if he does not elevate his hands. If he moves his hands under the water, in any way he pleases, his head will rise so high as to allow him free liberty to breathe; and if he will use his legs as in walking (or rather as if walking up stairs) his shoulders will rise above the water, so that he may use the less exertion with his hands, or apply them to other purposes. These plain directions are recommended to the recollection of those who have not learned to swim in their youth, as they may be found highly advantageous in preserving life.

COFFEE SYRUP.—This confection is exceedingly handy to travelers when proceeding on a long journey. Take half a pound of the best roasted ground coffee; boil the same in a saucepan containing three quarts of water, until the quantity is reduced to one quart; strain the water off, and, when refined of all impurities, introduce the liquor into another saucepan, and let it boil over again, adding as much Lisbon sugar as will make it a thick syrup, like treacle; remove it from the fire, and, when cold, pour it into bottles, corking the same down tight for use. Two teaspoonsful of the syrup introduced into a moderate sized tea cup, and filled with boiling water, will be fit for immediate use. If milk is at hand, use it *ad libitum*.

TO DRIVE AWAY RATS.—Take potash and wrap it in cotton batting, and place it in their holes; then stop them up. The rats will take the cotton to build nests with, and in so doing they will get their feet burnt with the potash, which will make them quit the premises.

GATHERING THE PERFUME OF FLOWERS.—The perfume of flowers may be gathered in a very simple manner, and without apparatus. Gather the flowers with as little stalk as possible, and place them in a jar three parts full of olive or almond oil. After being in the oil twenty-four hours, put them into a coarse cloth, and squeeze the oil from them. This process, with fresh flowers, is to be repeated according to the strength of the perfume desired. The oil being thus thoroughly perfumed with the volatile principle of the flowers, is to be mixed with an equal quantity of pure rectified spirits, and shaken every day for a fortnight, when it may be poured off, ready for use. As the season for sweet-scented blooms is now approaching, this method may be practically tested.

DIFFERENCE OF TIME.—The following table will show the difference of time (omiting seconds) between New York city and the principal cities of this country:

12 m.	New York is	12 17	p. m.	Portland, Me.
12 m.	"	12 12	p. m.	Boston, Mass.
12 m.	"	11 55	a. m.	Philadelphia.
12 m.	"	11 50	"	Baltimore.
12 m.	"	11 46	"	Richmond.
12 m.	"	11 44	"	Buffalo.
12 m.	"	11 37	"	Charleston.
12 m.	"	11 35	"	Pittsburg.
12 m.	"	11 33	"	Wheeling.
12 m.	"	11 30	"	Cleveland.
12 m.	"	11 29	"	Augusta, Ga.
12 m.	"	11 24	"	Detroit.
12 m.	"	11 24	"	Columbus.
12 m.	"	11 19	"	Cincinnati.
12 m.	"	11 14	"	Indianapolis.
12 m.	"	11 14	"	Louisville.
12 m.	"	11 06	"	Chicago.
12 m.	"	10 56	"	New Orleans.
12 m.	"	10 55	"	St. Louis.

12 m. New York is 10 44 a. m. St Paul.
12 m.   "   8 45   "   San Francisco.

The difference of time between New York and London is 4 hours and 55 minutes.

DOMESTIC DEPARTMENT.

DOMESTIC COOKERY.

BREAD, BREAKFAST AND TEA CAKES.—*Premium Bread.*— Salt or milk rising, one teacup new milk, and one teaspoon salt, pour in two teacups of boiling water; when cooled so as not to scald, stir the flour to make a batter, and set it in a kettle of warm water until it rises up light, which will be in about five hours; pour the latter into your pan of flour, and mix with warm water or milk in sufficiency to make four loaves of bread; add a teaspoonful of soda to the wetting, knead thoroughly, and put it in pans to rise, which it will do in half an hour.

Prize Corn Bread Recipe.—The prize of $10 offered by Orange Judd, the publisher of the "American Agriculturist," for the best corn bread loaf, was awarded to Mrs. James O'Brien, of Carrick, Pa. The recipe for making the bread is as follows: To two quarts of meal add one pint of bread sponge; water sufficient to wet the whole; add half a pint of flour and a tablespoonful of salt; let it rise; then knead well for the second time, and place the dough in the oven and allow it to bake an hour and a half.

Sir Charles Corn Bread.—Beat two eggs very light; mix with them one quart of Indian meal and one pint of sour milk or buttermilk. Add one tablespoonful of melted butter, and a teaspoonful of salæratus, well dissolved, immediately before baking. Beat hard and bake quick.

Worth Trying.—In times of high prices, the following recipe for making cheap bread, which we find in Hall's "Journal of Health," may be of service to some of our readers:

To two quarts of corn meal add one pint of bread sponge, water sufficient to wet the whole; add one half pint of flour and a teaspoonful of salt. Let it rise, then knead well, unsparingly for the second time. It costs just one half as much as bread from the finest family flour, is lighter on the stomach, and imparts more health, vigor and strength to the body, brain and bone. Three pounds of such bread (at five cents a pound for the meal) affords as much nutriment as nine pounds of good roast beef (cost, at twenty-five cents, $2.25) according to standard physiological facts.

Excelsior Corn Bread.—Seven cups corn meal, seven cups buttermilk, one cup wheat flour, one-half cup molasses, one teaspoon salæratus.

Corn Bread.—Take two quarts milk, five eggs, half ounce salt, one teaspoonful salæratus, one tablespoonful cream tartar, two ounces butter, and take Indian meal sufficient to make a thick batter; put into pans well greased, and then bake in a quick oven.

Corn Griddle Cakes.—Turn three pints of scalding milk to one quart meal, four tablespoonsful flour; when milk-warm, add four eggs, and a little salt. Bake on a griddle. If too thick, put in another egg and a little more milk.

Corn Griddle Cakes with Yeast.—Three coffee cups sifted Indian meal, one coffee cup flour, two tablespoons yeast, one saltspoon salt. Wet at night with water until as thick as pancakes. In the morning add one teaspoon soda dissolved.

To make Loaf Bread, etc.—To make loaf bread, hot breakfast cakes, and buckwheat cakes, superior to any thing of the kind before known, mix dry, and well rubbed together, two teaspoonsful of cream of tartar with one quart of flour; then dissolve three-fourths of a teaspoonful of super-carbonate of soda in a sufficient quantity of sweet milk; mix the whole together and bake *immediately*. If water be used instead of sweet milk, add a little shortening. If the above directions be strictly followed, bread will be produced of superior lightness and whiteness.

Buckwheats with Yeast. — One quart buckwheat flour, one teaspoon salt. Stir in warm water to make a thin batter. Beat thoroughly. Four tablespoons home-brewed yeast. Set the batter in a warm place. Let it rise over night. Add one teaspoon soda in the morning. Two tablespoons molasses.

Buckwheat Cakes without Yeast.—One quart buckwheat; one teaspoonful carbonate of soda dissolved in water sufficient to make a batter; when mixed, dissolve a teaspoonful of tartaric acid in hot water, mix well and bake immediately.

Buckwheat Cakes.—Mix a quart of buckwheat flour with a pint of luke-warm milk (water will do, but is not as good) and a teacup of yeast—set in a warm place to rise. When light (which will be in the course of eight or ten hours, if family yeast is used—if brewer's yeast is used they will rise much quicker), add a teaspoonful of salt—if sour, the same quantity of saleratus, dissolved in a little milk, and strained. If they are too thick, thin them with cold milk and water. Fry them in just fat enough to prevent their sticking to the frying-pan.

Crackers. — One pint water, one teacup butter, one tablespoon sour cream, one half teaspoon saleratus. Mix very hard, and pound well with the rolling pin.

Rye-and-Indian Bread. — Two quarts of Indian meal, scald it; add as much rye meal, one teacup molasses, one half pint lively yeast. Let it stand till it rises, from one to two hours. Bake it three hours.

Rye Drop Cakes.—One cup Indian meal, one cup rye meal, one half cup yeast, a little molasses and salt. Let them rise; add a little soda, if necessary.

Scarborough Puffs.—Boil one pint of new milk. Take out a cupful, and stir in flour to make a thick batter. Pour this into the boiling milk. Stir and boil till the whole is thick enough to hold a silver spoon standing upright. Take it from the fire, and stir in six eggs, one by one; add one teaspoon salt, and a dessert spoon of butter. Drop them by the spoonful into boiling lard, and fry like doughnuts. Grate sugar on them when fried.

French Tea Cakes.—To one pound of flour add two ounces of fresh butter; rub them together, then mix them with about four tablespoonsful of warm milk, one of beer yeast, and a beaten egg. Mix well together, and set the dough before the fire to rise. When it has risen, make it into three cakes, put them on buttered tins and place before the fire for an hour; then bake in a quick oven a quarter of an hour.

Excellent Fritters.—Boil two mealy potatoes, peel them, and rub them with two tablespoonsful of flour; peel and chop fine four sharp apples, and mix the whole into a batter with the beaten yolks of three, and the whites of two eggs; grate in a little nutmeg and ginger, and fry them in a pan of boiling lard.

Seed Cakes.—Four cups of flour, one and a half cups of cream or milk, one half cup butter, three eggs, one half cup caraway seeds, one teaspoonful salæratus, one teaspoonful rosewater; make it into a stiff paste, and cut them out with a tumbler. Bake twenty minutes.

Spice Snaps.—Take one and a quarter pounds of flour, half pound sugar, half pint molasses, six ounces of butter, half ounce of ginger, half ounce allspice: mix together; make in small pieces about the size of a marble; grease the pans well, and bake in cold oven.

Muffins.—One pint milk, sufficient flour to make a stiff butter, a tablespoonful of yeast, a little salt; let it rise, bake in rings on a hot griddle, or in a hot oven.

Dough Nuts.—Everybody and his wife, and particularly

his little folks, love the good old fashioned dough-nuts, or nut-cakes, or whatever name you choose to call them. But many persons are troubled with weak digestion (dyspepsia), and the large amount of grease absorbed by the said dough nuts does not always set well, but produces a rising in the stomach. When this is the case, try the following invention: The dough nuts being prepared as usual, just before immersing them into hot fat, plump them into a well-beaten egg. This will give them a thin coating of albumen, which will keep out the grease effectually. Furthermore, this coating retains the moisture, and keeps them in good condition much longer.

Another Recipe.—One teacup of sour cream or milk, two teacups of sugar, one teacup of butter, four eggs and one nutmeg, one teaspoonful of saleratus, flour enough to roll, cut into diamond cakes, and boil in hot lard.

Still another.—One cup sugar, two cups milk, one teaspoonful saleratus, flour.

Raised Doughnuts.—Two cups milk, two cups sugar, one cup butter, three eggs, nutmeg and cinnamon, one teacup yeast, as much bour as thought best. Raised until light.

Sponge Biscuit.—Stir into a pint of lukewarm milk half a teacup of melted butter, a teaspoonful of salt, half a teacup of family or a tablespoonful of brewer's yeast (the latter is the best); add flour till it is a very stiff batter. When light, drop this mixture by the large spoonful on to flat buttered tins several inches apart. Let them remain a few minutes before baking. Bake them in a quick oven till they are a light brown.

Cream Cakes.—Mix half a pint of thick cream with the same quantity of milk, four eggs, and flour to render them just stiff enough to drop on buttered tins. They should be dropped by the large spoonful several inches apart, and baked in a quick oven.

Rice Cakes.—Mix a pint of rice boiled soft with a pint of milk, a teaspoonful of salt, and three eggs beaten to a froth.

Stir in rice or wheat flour till of the right consistency to fry. If you like them baked, add two more eggs, and enough more flour to make them stiff enough to roll out, and cut them into cakes.

Crullers.—Take four pounds wheat flour, half pound butter, four eggs, one quart milk, one and a half pounds sugar, a little ground mace or nutmeg, and one ounce carbonate soda. Fry them in lard.

Another Recipe.—Two spoonsful white sugar, two spoonsful melted butter, two eggs, one half teaspoonful salæratus dissolved in a teacupful of milk; nutmeg and flour.

Molasses Cookies.—One cup molasses, one half cup butter, one half cup water, two teaspoonsful salæratus, one spoonful ginger, a little salt, and flour.

Bread Dough Cake.—Three cups of dough, three cups of sugar, one and a half cups of butter, four eggs, one teaspoonful salæratus, raisins, brandy.

Ginger Snaps.—One cup butter, one cup sugar, one cup molasses, one half cup ginger, one teaspoonful salæratus, and flour to make them hard.

Soft Gingerbread.—One pint molasses, one tablespoonful butter, one heaping teaspoonful of salæratus, one teaspoonful alum.

Indian, or Yankee Johnny Cake.—One pint milk, one pint meal, one tablespoonful molasses, one teaspoonful salæratus, salt and flour.

Cold Custard.—One quart new milk, one half pint cream, four ounces white sugar, one glass of white wine in which an inch of washed rennet has been soaked, nutmeg.

Lemon Jelly.—Take the clear juice of twelve lemons, one pound fine loaf sugar, one quart water. For each quart of the mixture put in an ounce of clarified isinglass. Let it boil up, and then strain into moulds.

To Clarify Isinglass.—Dissolve an ounce of isinglass in a cup of boiling water. Take off the scum and drain through a coarse cloth.

Pastry for Tarts.—One quart flour, one half pound butter, a little salt, two beaten eggs; add a little cold water; make it into a paste, set it away to cool, then roll it out again.

Cocoanut Cake.—One pound sugar, three-fourths pound flour, one half pound butter, five eggs; beat the yolks of the eggs with the sugar; the whites beat to a froth; beat the flour and butter together, then add the cocoanut, soda, and cream tartar. One good sized cocoanut, or two small ones; one half teaspoonful soda, one half teaspoonful cream tartar. Wash the cocoanut in the milk.

To make good Waffles.—One quart of milk (sour milk is best), one egg, one teaspoonful salæratus, one tablespoonful butter, one tablespoonful sugar. Flour to make it the consistency of batter for griddles.

Cream Tartar Cake.—One cup sugar, one half cup butter, one cup sweet milk, two eggs, two teaspoonsful cream tartar, one teaspoonful soda, and flour to make it the consistency of sponge cake.

A Good Common Cake.—One cup butter, two cups sugar, two eggs, one cup milk, one teaspoonful salæratus, nutmeg, flour, and raisins or currants.

Sugar Cookies.—One and a half cups sugar, one half cup milk, three fourths cup of butter, half teaspoonful salæratus, and one egg.

Shrewsberry Cake.—Make a stiff paste of a pound and a half of flour, three-quarters of a pound of sifted loaf sugar, a teaspoonful of pounded cinnamon, half a pound of warmed butter, one egg, and a little milk; roll it thin, cut and bake on tins in a quick oven.

Corn Muffins.—One pint of milk, two cups of Indian meal, two of flour, two eggs, a piece of butter (melted) the size of an egg, two teaspoonsful cream of tartar and one of soda—mix and bake on a hot griddle in muffin rings, turning them, so as to bake on both sides.

French Rolls.—Take a spoonful of lard or butter, three pints of flour, a cup of yeast, and as much milk as will work it up to the stiffness of bread. Just before you take them from the oven, take a clean towel and wipe them over with milk.

Premium Rusk.—Take one cup of sugar, a piece of butter (melted) the size of two eggs, six cups of flour, two cups of milk, four teaspoonsful cream of tartar and two of soda, and two eggs—mix and bake immediately in a hot oven. Eat hot. A beautiful rusk.

Barrington Rusk.—One cup sugar, one cup milk, one cup yeast, one cup flour. Mix over night. In the morning add one half cup sugar and one half cup butter rubbed together, two eggs, reserving the white of one, beaten to a stiff froth with a little sugar, to spread over the tops.

Almond Cake.—One pound of sugar, three-fourths of a pound of butter, three-fourths of a pound of flour, three ounces of sweet almonds and one bitter almond, (the almonds to be blanched, or put into hot water until the skin comes off, and then pounded or rolled), and moistened with two teaspoonsful of rose water—mix with the whites of seventeen eggs, and bake quick.

Delicate Cake.—One cup of butter, two cups sugar, one cup sweet milk, whites of seven eggs, four cups flour, two teaspoonsful cream of tartar, and one of soda.

Jelly Cake.—One cup of sugar, one of flour, three eggs, two teaspoonsful of cream of tartar, one of soda, a pinch of salt. Season with nutmeg, and stir well. Grease your dripping-pan, and spread the batter on even; bake ten minutes. When done, spread on the jelly when just from the oven. Begin at one end and roll it up, and wrap it in a cloth; when cool, it is fit to use. Cut in slices three-fourths of an inch thick.

Cup Cake.—One cup of butter, two cups of sugar, three cups of flour and four eggs, well beaten together, and baked in pans or cups. Bake twenty minutes, and no more.

Wheat Muffins.—One pint milk, two eggs, one tablespoon yeast, and one saltspoon salt. Flour to make a thick batter. Let it rise four or five hours, and bake in muffin rings. This can be made of Graham flour, with the addition of two tablespoons molasses, and is very nice.

Fruit Cake without Eggs.—Two pounds of flour, one and three-quarters pound of sugar, one half pound of butter, one pint of milk, one teaspoon salt, one and a half teaspoons soda dissolved in a little water, one nutmeg, one pound of raisins. This makes three loaves. Warm the milk, and add the butter and sugar beaten to a cream; then add the other ingredients.

Sponge Gingerbread.—One cup sour milk, one cup molasses, one half cup butter, two eggs, one and a half teaspoons salæratus, one tablespoon ginger. Flour to make as thick as pound cake. Warm the butter, molasses and ginger, then add the milk, flour and salæratus, and bake as quickly as you can.

Composition Cake.—One and three-quarters pound flour, one and one-quarter pounds sugar, three-quarters pound butter, one pound fruit, one pint milk, one teaspoon soda, four eggs, one glass brandy or wine, all sorts of spices. Bake in loaves.

Sugar Snaps.—One cup butter, two cups sugar, three eggs, one teaspoon soda, one tablespoon ginger. Flour to roll.

Ginger Snaps.—One cup molasses, one half cup sugar, one half cup butter, one half cup warm water, the butter melted with it, and one teaspoon soda dissolved in it; two tablespoons ginger. Flour to make stiff. Knead it well; roll and cut in round cakes, and bake in a moderate oven.

Loaf Cake.—Two pounds dried and sifted flour, one pint new milk blood-warm, one-quarter pound of butter, three-quarters pound of sugar, one pint home-brewed yeast, three eggs, one pound stoned raisins, one nutmeg, a glass of wine if you like. Rub the butter and sugar to a cream, add the flour. Add the other ingredients, and let it rise over night. Bake one hour and a half in a slow oven.

Beautiful Sponge Cake.—Twelve eggs beaten separately,

one pound of flour, and three-quarters of a pound of sugar —mix the sugar and yolk of eggs first, then add the whites (beaten to a froth), and lastly, stir in the flour gradually. Use no soda, but flavor with any thing you like.

Another.—Six eggs, two teacupsful of sugar, one and a half of flour, one teaspoonful of cream of tartar, one half teaspoonful of soda, and one teaspoonful essence of lemon; beat the whites of the eggs till very light; mix the yolks with the sugar, beat till very smooth; mix the soda and cream of tartar with the flour, then add to the former mixture; then add the lemon. The whole should be stirred slowly till the top is covered with bubbles. Bake in a quick oven.

Crullers.—Three eggs, one teacup of sugar, a little salt, a piece of butter the size of an egg; knead in flour enough to make a stiff dough. Roll out about half an inch thick; cut in strips an inch wide and six inches long, then slit one edge half through to once in half an inch, then join the ends round, and fry in hot lard to a nice brown; put away in jars; they will keep for months.

Marble Cake—for the white cake—One cup butter, three cups white sugar, five cups flour even full, one half cup sweet milk, one half teaspoon soda, whites of eight eggs; flavor with lemon.

For the dark cake—One cup butter, two cups brown sugar, one cup molasses, one cup sour milk, one teaspoon soda, four cups flour, yolks of eight eggs, and one whole egg; spices of all sorts. Put in pans first a layer of dark cake, then a layer of the white, and so on, finishing with a layer of dark cake. Bake in a hot oven.

Cookies—One cup of butter, two cups of sugar, one cup of cold water, half a teaspoonful of saleratus, two eggs, flour enough to roll, and no more.

Cheap Fruit Cake.—To one quart of sifted flour, add a teacup of sugar, a half a cup of butter, two teaspoonsful cream tartar, one of soda; rub them all thoroughly together into the flour; stir in cold water sufficient to make a stiff batter; pour it into a small tin pan; bake one hour—in a quick oven the first half hour, then quite slow; spice with any kind to suit the taste, and add a teacup of raisins.

Cheap Tea-cake.—One cup of sugar, one of butter (or sour milk), one of sour cream, one teaspoonful of soda, one egg, a little salt and nutmeg; stir in flour to make a stiff batter; bake until done.

Mount Pleasant Cake.—Two and a half cups sugar, five cups flour, one cup butter, one cup sour milk, one teaspoon soda, four eggs. Mace, citron and currants to taste.

Elmsdale Cake.—Six cups of sugar, three cups of butter, two cups of buttermilk, seven eggs, one teaspoonful of salæratus, ten cups of flour, some raisins, one nutmeg. Rub the sugar and butter together, add the buttermilk and eggs, and put in the salæratus the last thing. This is very much like pound cake, only not quite so rich.

Nice Doughnuts.—One pint sour milk, soda enough to sweeten, one teacup of sugar, seven tablespoonsful of melted lard, two eggs, flour enough to mix up soft.

Citron Cake.—One teacup sugar, two-thirds teacup butter two teacups flour, one half teacup milk, and one teaspoon soda dissolved in it. Flavor with essence lemon, and put in small bits of citron cut thin. Bake in hearts and rounds.

Caraway Cakes.—Two quarts flour, one cup butter, one quart sugar after it is rolled, one half pint of caraway seeds, little essence lemon, one half teaspoon soda dissolved in water. Roll out and bake in thin cakes.

Honey Cake.—One quart strained honey, one half pint sugar, one half pint melted butter, one teaspoon soda dissolved in one half teacup warm water, one half a nutmeg, and one teaspoon ginger. Mix these ingredients, and then work in flour to roll. Cut in thin cakes, and bake on buttered tins in a quick oven.

Seed Cake.—Four cups of flour, one and a half cup of cream or milk, half a cup of butter, three eggs, half a teacupful of caraway seeds, a teaspoonful of salæratus, the same of rose-water; make it into a stiff paste, and cut them out with a tumbler or biscuit-cutter. Bake about twenty minutes.

Chocolate Puffs.—Beat stiff the whites of two eggs, and beat in gradually one half pound powdered sugar; scrape fine one ounce and a half prepared cocoa, dredge it with flour, mixing in the flour well; add this gradually to the eggs and sugar; stir the whole very hard. Cover the bottom of a pan with a sheet of white paper; place on it thin spots of powdered sugar about the size of a half dollar. Pile a portion of the chocolate mixture on top of each, smoothing with a knife wet in cold water, and sift a little sugar over each. Bake in a quick oven a few minutes. When cold, loosen them from the paper with a broad knife.

Ginger Drop Cake.—One cup molasses, one cup lukewarm water, one half cup shortening, one half tablespoonful ginger, one teaspoon soda, flour to make a stiff batter. Beat fifteen minutes, and drop on buttered pans.

One-egg Tea-cake.—One egg, four tablespoonsful of white sugar, one tablespoonful of butter, a gill of milk, one teaspoonful of yeast powder, and enough flour to make it the consistency of pound cake; season to taste. Bake in patties.

Fruit Cake.—One and a half pounds of sugar, one and a quarter pounds of flour, three-quarters of a pound of butter, six eggs, a pint of sweet milk, one teaspoonful of salæratus, one glass of wine, one of brandy, and as much fruit and spice as you can afford.

A Chicago Sponge Cake.—Ten eggs, one pound sugar, half pound of flour, and lemon juice or extract to flavor. Beat the whites to a stiff froth, warm and sift the flour; stir the yolks and the sugar together, till light, and add the whites and flour, half at a time, alternately. Stir the whole gently, till bubbles rise to the surface. Bake in a moderate oven.

Soft Gingerbread.—Two cups of molasses, one cup of sugar

one of butter, one of sour milk, and five of flour; three eggs, one teaspoonful of salæratus, and a tablespoonful of ginger. All the materials should be warmed before being mixed.

In making gingerbread with sorghum molasses, mix the soda with the molasses; then warm, stir till light, then mix with flour in the usual way, which will make light bread.

Boston Fruit Cake.—Three-quarters of a pound of butter, three-quarters of a pound of sugar, three-quarters of a pound of flour, eight eggs, one gill of cream, one teaspoonful of cinnamon and nutmeg mixed, half a gill of brandy, one pound of currants (washed, dried and picked), one pound of raisins (seeded and chopped); beat the butter, sugar and spice until very light, then stir in the cream and one-fourth of the flour; whisk the eggs until thick, which add by degrees; then the remainder of the flour, half at a time; lastly the fruit; beat all well together. Butter and line your pan with white paper, and bake in a moderate oven.

Gold Cake.—One cup butter, one cup milk, two cups sugar, three cups flour, yolks of five eggs, one small teaspoonful soda, two teaspoonsful cream tartar; flavor with nutmeg or vanilla.

Silver Cake.—One half cup butter, one half cup sugar, one cup milk, two and one half cups flour, whites of five eggs, one teaspoonful cream tartar, one half teaspoonful soda.

SODA CRACKERS.—Take one cupful of shortening, two teaspoonsful of cream tartar, and rub them in ten cupsful of flour. Afterward add one cupful of sweet milk, one of water, one teaspoonful of soda, and a little salt. Knead until the dough is smooth. Roll thin, cut in squares, and bake quick.

TO MAKE FLOUR CAUDLE.—Into five spoonsful of the purest water, rub smooth one dessertspoonful of fine flour. Set over the fire five tablespoonsful of new milk, and put two bits of sugar into it: the moment it boils, pour into it flour and water, and stir it over a slow fire twenty minutes. It is a nourishing and gentle stringent food, particularly for babies who have weak bowels.

BAKED TOMATOES.—After removing the skin by pouring boiling water over them, cut the large ones in two or more pieces, and put them in a baking dish, and season by using salt, pepper, sugar, and butter. A little flour and water should be added, and they should be baked in a hot oven for an hour, when they will be found delicious and healthful.

DESSERT DISHES.—*Rock Cream.*—Boil a teacupful of the best rice till quite soft, in new milk, sweeten it with powdered loaf sugar, and pile it up on a dish. Lay on it in different places square lumps of either currant jelly or preserved fruit of any kind; beat up the whites of five eggs, with a little powdered sugar, and flavor with orange-flower water or vanilla. Add to this, when beaten very stiff, about a tablespoonful of rich cream, and drop it over the ice, giving it the form of a rock of snow.

Ornamental Froth.—Beat the whites of four eggs to a froth, then stir in half a pound of preserved strawberries, or raspberries; beat the whole well together, then turn it over the top of blanc mange or creams.

Pink Ice-cream.—Take three gills of currant or strawberry juice; mix with half a pound of white powdered sugar a pint and a half of thick cream. Whisk it till well mixed, serve it up in a glass dish, or freeze it if you like.

A Dish of Snow.—Grate a cocoanut, leaving out the brown part, heap it up in the centre of a dessert dish, ornament it with myrtle, or box; serve it up with snow cream, or cream and white sugar.

Snow Cream. Beat the whites of five eggs to a stiff froth, then stir in two large spoonsful of powdered white sugar, a large spoonful of sweet wine, and a teaspoonful of rose-water. Beat the whole well together, then add a pint of thick cream. This is a nice accompaniment to a dessert of sweetmeats.

Essence of Nutmeg.—Dissolve an ounce of the essential oil of nutmeg in a pint of rectified spirits. This is very nice to use for flavoring cakes and puddings.

TARTS.—*Marlborough Tart.*—Quarter and stew tart apples till soft enough to strain through a sieve. To twelve large spoonsful of the strained apple put three of melted butter, the juice and grated rind of a lemon, half a nutmeg, half a pint of milk, a wine-glass of wine, four beaten eggs, and sugar to the taste. Bake the tarts with a lining and rim of pastry. Ornament them with small strips of the pastry.

Quince Tarts.—Stew and strain six quinces, mix with them half a pound of sugar, half a pint of cream, four eggs, and nutmeg to the taste.

PUDDINGS.—*Gingerbread Pudding.*—One half pint molasses, one half pound flour, one half pound chopped suet, two tablespoons ginger. Boil four hours. Liquid sauce.

Corn Starch Pudding.—One quart of sweet milk brought to boil, and a little salt, two eggs well beaten, three heaping teaspoonsful of corn starch, with the addition of a little sweet milk. Stir well. It will cook in four or five minutes. Serve with sweetened cream.

Apple Sago Pudding.—Core six apples, fill them with sugar; lay them in a pudding dish; pour over them one cup of sago, previously soaked in water one hour. Flavor with lemon if the apples are not tart. Bake one hour and a half.

Boiled Plum Pudding.—Take one pound of good suet; cut it in small pieces, and add one pound of currants, and one of stoned raisins, eight eggs, one nutmeg, grated, one pound of flour, and one pint of milk; to the eggs, previously well beaten, add one half the milk, and mix well together; stir in the flour, spice, fruit and suet, and the remainder of the milk. Boil from four to five hours.

Indian Meal Pudding.—Into one quart of boiling milk stir one quart of sifted fine meal, then add one quart of cold milk, two well-beaten eggs, one half teacupful of sugar, one teacupful of flour, and a little salt and spice to suit your taste; stir it well, and pour it into a buttered dish, and bake two hours. Serve with butter.

Potato Pudding.—Two pounds potatoes boiled and mashed, one half pound sugar, one half pound butter, six eggs, wine-glass brandy, one nutmeg. Line a dish with paste, and bake.

Baked Apple Pudding. — Pare and quarter four large apples; boil them tender with the rind of a lemon in so little water that when done no water may remain; beat them quite fine in a mortar; add the crumbs of a small roll, quarter pound of butter, melted, the yolks of five and whites of three eggs, juice of a half lemon, sugar to your taste; beat all well together, and bake in paste.

Old English Plum Pudding.—To make what is termed a pound pudding, take of raisins well stoned, currants thoroughly washed, one pound each; chop a pound of suet very finely, and mix with them; add a quarter of a pound of flour, or bread very finely crumbled, three ounces of sugar one ounce and a half of grated lemon peel, a blade of mace, half a small nutmeg, half a dozen eggs well beaten; work it well together, put it into a cloth, tie it firmly—allowing room to swell—and boil not less than five hours. It should not be suffered to stop boiling.

Plain Steamed Pudding.—One quart buttermilk, one heaping teaspoon soda, a little salt, flour to make a stiff batter. Steam one hour and a half. Liquid sauce.

Steamed Wheat Flour Pudding.—One quart of sour milk, half a teacupful of sour cream, two eggs, one teaspoonful of soda, and a little salt; stir in flour, so as to make a stiff batter, steam one hour, and serve with sweetened cream.

Lizzie S.'s Pudding Sauce.—One cup sugar, piece of butter size of an egg, one glass wine. Beat the sugar and butter to a cream; add one egg well beaten, and one half wine glass boiling water, the last thing before serving.

PIES.—*Summer Mince Pies.*—Four crackers, one cup and and a half sugar, one cup molasses, one cup cider, one cup water, two-thirds cup butter, one cup chopped raisins, two eggs beaten and stirred in last thing. Brandy and spice to taste.

Lemon Cream Pie.—One cup sugar, one cup water, one raw potato grated, juice and grated rind of one lemon. Bake in pastry top and bottom. This makes one pie.

Cocoanut Pie.—Two grated cocoanuts, four eggs, two and a half cups sugar, three cups milk, piece of butter size of an egg.

Grape Pie.—Grapes are the best for pies when tender and green; if not very small they require stewing and straining to get rid of the seeds; sweeten to the taste when strained. If the pies are made of whole grapes, allow half a pint of sugar to a medium sized pie; put in the sugar and grapes in alternate layers, in deep pie-plates; add a tablespoonful of water to each pie.

Ripe Cherry Pie.—Remove the stems and stones from the cherries, cover the bottom of a long tin with the fruit, to which add a teacupful of sugar and one of flour; bake with an upper and under crust.

Mince Pies.—To make mince pies without apples or cider, take the requisite quantity of meat, and one-third the quantity of beets that is commonly used of apples. Boil the beets, and let them pickle twelve hours. Chop them very fine and add one-eighth of grated wheat bread. Sweeten and season with spices, etc., to taste.

Apple Mince Pies.—To one dozen common sized apples

chopped fine, add six eggs well beaten, and half a pint of cream. Put in spices and sugar, and raisins or currants, just as you do in meat mince pies.

Lemon Pie.—Take one lemon, grated entire, one cup of sugar, three eggs, one tablespoonful of flour, and two cups of cold water—mix, make, and bake like custard pie. Very good.

Custard Pie.—One pint of milk, two eggs, a tablespoonful of sugar, a little salt, and a little nutmeg, grated on. For crust use common pastry.

Rice Pie.—Boil the rice soft; and to each pie, put in one egg, one tablespoonful of sugar, a little salt, and a little nutmeg.

CUSTARD.—Allow four eggs to each pint of fresh milk. Reserve part of the whites to froth and lay on top. Beat the eggs smooth, and stir them in the milk; sweeten with best loaf sugar. Set a bucket with the mixture in a pot of boiling water. Stir until done, and remove from the fire instantly. The same mixture may be baked.

Apples or Quinces, peeled and cored, with the hole made by coring filled with jelly or brown sugar, and baked with a little wine and sugar around, are very nice; with a custard poured over and baked, they are termed "a bird's nest."

SIDE DISHES, ETC.—*Fricassee Chicken.*—Cut the chickens up; let it lie in water for an hour; dry them in a towel; then put them in a stewpan with just water enough to cover them, with a little mace, part of an onion cut up fine, and a little sweet marjoram. Boil them until tender. Then take one-fourth pound of butter, and rub some flour with it until perfectly smooth, and drop the butter and flour into the chicken water, stirring it all the time until it boils. Then beat up the yolk of an egg with a little cream, and pour in when done.

A Nice Way to Cook Chickens.—Cut up the chicken, put it in a pan and cover it with water; let it stew as usual, and when done, make a thickening of cream and flour, adding a piece of butter, and pepper and salt; have made and baked a pair of short-cakes, made as for pie-crust, but roll thin, and cut in small squares. This is much better than chicken pie, and more simple to make. The crusts should be laid on a dish, and the chicken gravy put over it while both are hot.

Delicious Dressing for Fowls.—Spread pieces of stale but tender wheaten bread liberally with butter, and season rather high with salt and pepper, working them into the butter; then dip the bread in wine, and use it in as large pieces as is convenient to stuff the bird. The delicious flavor which the wine gives is very penetrating, and it gives the fowl a rich, gamy character which is very pleasant.

To Fry Eggs.—Fry very slowly and but slightly—that is, softly and tenderly done, the yolk mostly soft. Fry in butter, if a rare dish is wanted, and let the butter be fresh. If the eggs are entirely immersed in the butter, all the better. It is a delicate operation, and requires but little heat, so that the butter is not hurt, only well melted.

To Warm up Cold Beefsteaks.—Put a fine minced onion in a stewpan, add half a dozen cloves and as many peppercorns, pour on a coffee cup of boiling water, and add three tablespoons butter. Let it simmer ten minutes. Then cut up the meat in pieces an inch square, and let it simmer in this gravy about five minutes. Three large tomatoes stewed with the onion improves this.

Sausage Meats.—Take one-third fat and two-thirds lean pork, and chop them. To every twelve pounds meat add twelve large spoons powdered salt, nine of sifted sage, and six of sifted black pepper. Add summer savory if you choose. Make in cakes, and heep in a cool, dry place.

Bologna Sausage.—Chop fine equal portions of veal, pork,

and ham, season with sweet herbs and pepper; put them in cases, boil them till tender, and dry them.

Another Recipe.—Chop fine and mix well together one pound each of beef suet, fresh pork, bacon, fat and lean, fresh beef, and veal; add a handful of sage leaves, powdered fine, with a few sweet herbs; season pretty high with pepper and salt. Take a large, well-cleaned gut and fill it. Set a saucepan on the fire with water. When it boils, put in the sausage, first pricking it to prevent its bursting. Boil it one hour.

Sausage Stew.—Put in a thick layer of peeled and sliced potatoes; add salt and a layer of sliced sausages; repeat till you fill your dish, having a layer of potatoes on top. Pour in a little water and some butter, and cook till done.

Beef or Veal Stew, with Apples.—Rub a stew pan with butter; cut the meat in thin slices, and put in with pepper, salt, and apple, sliced fine, onion, if you like; cover tight, and stew till tender.

MUTTON HARICOT.—Make a rich gravy by boiling the coarser parts for the liquor, seasoning with pepper, spice, and catsup. Cut into this gravy carrots, parsnips, onions, and celery, boiled tender. Then broil the mutton, seasoning it with salt and pepper; put it into the gravy, and stew about ten minutes.

SOUP.—Take a shin of beef, six large carrots, six large yellow onions, twelve turnips, one pound of rice or barley; parsley, leeks, summer savory; put all into a soup-kettle, and let it boil four hours; add pepper and salt to taste: serve all together.

French Vegetable Soup.—To a leg of lamb of moderate size, put four quarts water. Of potatoes, carrots, onions, tomatoes, cabbage, and turnips, add a teacup each, chopped fine; salt and pepper to taste. Let the lamb be boiled in this water. Let it cool; skim off all the fat that rises to the top. The next day boil again, adding the chopped vegetables; Let it boil three hours the second day.

Mock Turtle Soup.—Cut the meat clean from the bones of a fine calf's head. Then boil the bones in a quart of water until the liquor is reduced to a pint; season it with cayenne, nutmeg, and mace; pour into it a pint of Madeira wine, and a little parsley and thyme.

WELSH RABBIT.—Cut one pound of cheese into small slips, if soft—if hard, grate it down. Put it into a tin dish, with an ounce of butter, and set the dish over a spirit-lamp, or a gentle fire. Have ready the yolk of an egg, whipped with half a glass of Madeira wine, or as much ale or beer. Stir your cheese, when melted, until thoroughly mixed with the butter; then add gradually the egg and wine. Keep stirring it till it forms a smooth mass. Season with cayenne pepper and grated nutmeg. To be eaten with a thin, hot toast.

PICKLED OYSTERS.—To each quart of oyster liquor put one teaspoon pepper, two blades of mace, three tablespoons white wine, four tablespoons vinegar, one teaspoon allspice, one tablespoon salt. Simmer in this five minutes, then put the oysters in jars. Boil the pickle, skim it, and pour over them.

TO BOIL FISH.—Fill the fish with a stuffing of chopped salt pork and bread, seasoned with salt and pepper, and sew it up. Sew all in a cloth, or it will not take up well. Put it in cold water enough to cover it, salt it, and add three tablespoons vinegar. Boil it slowly for twenty or thirty minutes, and serve with drawn butter and caper sauce.

How Salt Fish should be Freshened.—Many persons who are in the habit of freshening mackerel or other salt fish, never dream that there is a right and a wrong way to do it. Any person who has seen the process of evaporation going on at the salt works, knows that the salt falls to the bottom. Just so it is in the pan where your mackerel or your whitefish lies soaking; and, as it lays with the skin side down, the salt will fall to the skin, and there remain, when, if placed with the flesh side down, the salt falls to the bottom of the pan, and the fish comes out freshened as it should be. In the other case, it is nearly as salt as when put in. If you do not believe this, test the matter for yourselves.

Good Way of Cooking Onions.—It is a good plan to boil onions in milk and water; it diminishes the strong taste of that vegetable. It is an excellent way of serving up onions, to chop them after they are boiled, and put them in a stewpan, with a little milk, butter, salt, and pepper, and let them stew about fifteen minutes. This gives them a fine flavor, and they can be served up very hot.

To Sprout Onions.—Pour hot water on the seed, and let it remain two or three seconds. They will sprout almost immediately, and should be planted without delay; if so, they will come up much earlier than seed that was not scalded.

Pepper Vinegar.—Get one dozen pods of pepper when ripe; take out the stems, and cut them in two; put them in a kettle, with three pints of vinegar; boil it away to one quart, and strain it through a sieve. A little of this is excellent in gravy of every kind, and gives a flavor greatly superior to black pepper; it is also very fine when added to each of the various catsups for fish sauce.

Good Tomato Catsup.—Wash and jam the tomatoes; then bring them to a boil, and strain through a sieve. To two quarts of the juice add one pint of good cider vinegar, and boil down one-third or one-half; then add cloves, spice, cinnamon, salt, cayenne and black pepper, to taste; then boil a few minutes, and when half cold, bottle for use.

Another Recipe.—To a gallon of skinned tomatoes add four

tablespoonsful salt, four tablespoonsful black pepper, half spoonful allspice, eight red peppers, and three spoonsful of mustard. All these ingredients must be ground fine, and simmered slowly in sharp vinegar for three or four hours. As much vinegar is to be used as to leave half a gallon of liquor when the process is over. Strain through a wire sieve, and bottle and seal from the air. This may be used in two weeks, but improves by age, and will keep several years.

Another.—To one gallon tomatoes add four tablespoons salt, four tablespoons cloves, one tablespoon mace, one tablespoon cayenne, two tablespoons allspice, eight tablespoons white mustard seed, two whole peppers, one ounce garlic, one pint good vinegar. Boil away nearly half. Strain and bottle. Cork tight.

Warning to Housekeepers.—George D. Armstrong, an old resident of the township of Lodi, died from eating tomato catsup, made in a copper boiler,—the acid of the fruit, by a chemical process, combining with the copper, forming what is called in chemistry disaconite of copper (common verdigris). We are informed that the catsup was allowed to remain in the boiler for some three days after being made, the result, no doubt, of a want of knowledge of the chemical affinity of the two substances—the acid of the tomato, and the copper of the boiler. Mr. Armstrong, being exceedingly fond of the tomato in that form, disregarding the unpleasant taste that other members of the family complained of, ate bountifully of it, resulting in death after a number of days' terrible suffering.

The Tomato as Food.—A good medical authority ascribes to the tomato the following very important medical qualities:

1st. That the tomato is one of the most powerful aperients of the liver and other organs; where calomel is indicated, it is one of the most effective and the least harmful medical agents known to the profession.

2d. That a chemical extract will be obtained from it that will supersede the use of calomel in the cure of diseases.

3d. That he has successfully treated diarrhœa with this article alone.

4th. That when used as an article of diet, it is almost sovereign for dyspepsia and indigestion.

5th. That it should be constantly used for daily food. Either cooked or raw, or in the form of catsup, it is the most healthy article now in use.

YEAST.—Boil one pound of good flour, a quarter of a pound of brown sugar, and a little salt, in two gallons of water, for one hour. When milk-warm, bottle it and cork it close. It will be fit for use in twenty-four hours. One pint of this yeast will make eighteen pounds of bread.

Potato Yeast.—Six potatoes boiled and mashed, one cup flour, one half cup sugar, tablespoon salt. Turn to this, one pint boiling water, then one pint cold water. Raise it with a cup of yeast. Set it in a warm place, and it will rise frothing in a few hours. It is now ready for use. Set it in a cool place. It will keep only a few days.

COFFEE.—*How to Settle it.*—A common method of clearing coffee is by the addition of an egg. The white is the only valuable part for the purpose, and only a small portion of one is needed for an ordinary family. It should be mixed with the ground coffee before the water is added. Clean egg-shells will do very well. When eggs are fifty cents a dozen, they are not always at hand; a bit of codfish, or even a pinch of salt, is a very good substitute; and if the coffee is put to soaking in a little cold water over night, it will settle clear without the addition of anything.

Acorn Coffee (a pleasant beverage).—Take sound, ripe acorns, peel off the hull or husk, divide the kernels, dry them gradually, and then roast them in a close vessel. When roasted, add a little butter in small pieces, while hot, in the roaster. Grind like other coffee, and to each teacupful add a tablespoonful of common coffee. To be made and drank as common coffee.

Never keep coffee long in a tin coffee pot, as the vessel will impart its flavor to the fluid—or rather the fluid will abstract it.

Keep coffee where it will not imbibe odor, for it takes it on readily. Whole cargoes are sometimes lost by the presence of allspice or rum.

To make Coffee.—The best mode is the French mode of racking—that is, to strain it. You can get it clear—all the flavor and all the strength—and it is done as soon as hot water can run through it. French coffee pots for racking can be had at any house-furnishing store.

Coffee as the French Prepare it.—In Paris the coffee is nectar compared with the beverage we in common call coffee. The French use three kinds of coffee, Mocha, Java and Rio, mixed in equal parts. The coffee before roasting is winnowed, to cleanse it of the dust, etc.; it is then culled, or picked over: every black or defective kernel is picked out, as well as small stones, seeds, and rat droppings, which are abundant in most coffee. It is then put into tubs of clean water and well washed, then spread to dry, and when dry, is ready for roasting. The coffee required is roasted daily at large establishments; while warm, it is ground, and put up in papers of a conical shape, holding from two ounces to half a pound, and sealed up. Gentlemen, as they leave their places of business for home and dinner, when convenient, call and take the needed supply. The coffee is put into a pot or digestor in cold water, and then set over a lamp expressly for the purpose, and there heated—not boiled, but steeped; from this digestor no steam or fume arises; it is coffee, and a beverage delicious, health-giving; not the bitter, acid, filthy, nauseous drug we are in the habit of partaking of, and calling coffee.

A Substitute for Coffee.—Boil clean white rye until the

grains swell; then drain and dry it. Roast it to a dark brown, and prepare as other coffee, allowing twice the time for boiling. This alone makes good coffee, but if you add a little of the extract of coffee, or some beets or carrots sliced thin and dried in an oven till brown, it will make a coffee but little, if any, inferior to the genuine article.

Another.—Roast Indian meal in a bread-pan to a very dark brown; then mix it with molasses to a thick batter, and put it in a pan, and bake slowly until it becomes dry. It can be stirred often, so as to make it crummy, or baked in a cake, and a piece ground, or pounded fine, when wanted for use. This is said to make good coffee.

TEA.—*How to Draw it.*—Pour tepid or cold water enough on the tea to cover it, place it on the stove-hearth, top of a tea-kettle, or any place where it will be warm, but not enough so as to cause the aroma to escape in steam. Let it remain about half an hour, then pour on boiling water, and bring to the tea table.

Tea in Russia.—They drink tea, in Russia, as soon as the boiling water is poured on it, whilst we allow it to stand until it becomes as black as one's hat, and as bitter as hops. The men mostly drink their tea in tumblers, without milk, sometimes adding a slice of lemon, whilst the women take it in cups, with any amount of cream.

Native Tea.—Good meadow hay, one ounce; add boiling water, one quart. It is as much superior to the dried, poisoned and adulterated leaves of China, as gold is superior to lead. It is delicious to the palate—it is saccharine and aromatic—stills the nerves at night—is anti-bilious—promotes digestion, and gives appetite.

As a healthy drink, in place of tea, Dr. Thompson, in a late work of his, recommends the use of the dried leaves of the red raspberry. They cleanse the system of canker, and thus act beneficently to health. The leaves should be gathered in a good, dry chamber, on clean boards, or paper, to dry.

HARD BUTTER WITHOUT ICE.—To have delightfully hard butter in summer, without ice, put a trivet, or any open flat thing with legs, in a saucer; put on this trivet the plate of butter; fill the saucer with water, turn a common flower-pot upside down over the butter, so that its edge shall be within the saucer and under the water; plug the hole of the flower-pot with a cork, then drench the flower-pot with water; set it in a cool place until morning, or if done at breakfast, the butter will be very hard by supper time.

Preparing and Preserving Butter.—After the cream is placed in the churn, pour by small portions at a time, agitating the while, sufficient lime water to destroy the acidity. Churn until the butter is separated; it will not collect in lumps; pour off the buttermilk, and churn till it is all collected. More butter will be obtained, and will keep much longer.

Churning.—In churning butter, if small granules of butter appear which do not "gather," throw in a lump of butter, and it will form a nucleus, and the butter will "come."

How to Freshen Salt Butter.—Churn the butter with new milk, in the proportion of a pound of butter to a quart of milk. Treat the butter in all respects as if it was fresh. Bad butter may be improved greatly by dissolving it thoroughly in hot water. Let it cool, then skim it off and churn again, adding a small quantity of good salt and sugar. The water should be merely hot enough to melt the butter, or it will become oily.

To Pack Butter.—Pack your butter in a clean, scalded firkin, cover it with strong brine in which a bit of saltpetre is dissolved, spread a cloth all over the top, and it will keep well.

To Increase the Quantity of Butter.—While the milking of your cows is going on, let your pans be placed on a kettle of boiling water. Turn the milk into one of the pans taken from the kettle of boiling water, and cover the same with another of the hot pans, and proceed in the same manner with the whole mess of milk, and you will find that you have double the quantity of sweet and delicious butter.

To make Butter Yellow.—Just before the churning is completed, put in the yolks of eggs.

A Good Way to make Butter.—Put sweet milk into tin pans, and simmer on a stove for fifteen minutes, being careful not to burn the milk; then churn in the usual manner. In this way butter may be produced almost immediately, and of superior quality to that made in the usual way, and will keep sweet much longer.

A SUBSTITUTE FOR CREAM.—Cream, when unattainable, may be imitated thus:

Beat the white of an egg to froth, put in a small lump of butter, and mix well; then turn the coffee to it gradually, so that it may not curdle. If perfectly done it will be an excellent substitute for cream. For tea, omit the butter, using only the egg. This might be of great use at sea, as eggs can be preserved fresh in various ways.

ARTIFICIAL HONEY.—This recipe has been sold for thousands of dollars: One pint water, one eighth ounce alum; boil; set off; put in four pounds white sugar, boil one minute, strain; when milk-warm, add one teaspoonful of flavoring for artificial honey, made in the following manner: one half pint best alcohol, three drops ottar of roses, one half ounce Jamaica ginger, shake well: use in three days.

TO PRESERVE MILK.—Provide bottles, which must be perfectly clean, sweet and dry. Put the milk warm from the cow into these bottles, and as they are filled, immediately cork them up well, and fasten the corks with wire. Then spread a little straw on the bottom of a boiler, on which place the bottles, with straw between them. Fill it up with cold water. Heat the water, and as soon as it begins to boil, draw the fire, and let the whole gradually cool. When quite cold, take out the bottles, and pack them away in sawdust, and put them in a cool place, but where the milk will not freeze. Milk preserved in this way will keep perfectly sweet for years.

Another Method.—A little horse radish put into a pan of milk, will keep it sweet for several days, either in the open air or in the cellar.

FROZEN POTATOES.—If your potatoes freeze in the cellar, don't wait for them to thaw, but throw them into a conical heap, either where they are, or in the open air, and cover them with dirt, straw, shavings, old clothes or chaff, packed tight with them, and they are safe. The cover will prevent sudden changes, which cause all the mischief.

TO CLEAN KNIVES.—A small, clean potato, with the end cut off, is a very convenient medium of applying brick dust to knives, keeping it about the right moisture, while the juice of the potato assists in removing stains from the surface. A better polish can be obtained by this method than by any other we have tried, and with less labor.

SODA WATER.—Super-carbonate of soda, one-fourth ounce; tartaric acid, one-sixteenth ounce; sugar, one ounce; water, half pint.

Lemonade.—Tartaric acid, one ounce; sugar, three ounces; essence of lemon, one drachm; water, two quarts.

MAKING LARD.—Cut the fat up into pieces about two inches square; fill a vessel holding about three gallons with the pieces; put in a pint of boiled lye, made from oak or hickory ashes, and strain before using; boil gently over a slow fire, until the cracklings have turned brown; strain and set aside to cool. By the above process you will get more lard, a better article, and whiter than by any other process.

Candles from Lard.—Take one pound of lard, heat it quite hot, then add one cent's worth of aquafortis, and stir till well mixed. After standing a few minutes, a scum will rise; remove this, and then your lard will be fit for moulding into candles. They will be equally as hard as tallow, burn as well, and give as good a light. There is no unpleasant smell arising from the use of them. The kettle had best be put out in the yard when the aquafortis is put in, as the smell is at that time very disagreeable.

To Prevent Mildew on all sorts of trees, keep the plant subject to it occasionally syringed with a decoction of elder leaves, which will prevent the fungus growing on them.

Herbs for Winter Use.—June is the time to gather herbs for the coming winter. Some kinds will be wanted for the kitchen, and others for medicine. As they are coming into flower, they should be cut and hung in the shade until thoroughly dry, after which they may be put into coarse muslin bags, or closely wrapped in paper, and properly labeled. Keep them in a dry place, free from flies, dust, etc.

To dry Herbs.—Gather them on a dry day, just before they begin to blossom; brush off the dust, cut them in small branches, and dry them quickly in a moderate oven; pick off the leaves when dry, pound and sift them, bottle them immediately, and cork them closely. They must be kept in a dry place.

To Keep Moths From Furs, etc.—A piece of camphor placed at the bottom of a drawer of woolens or furs will prevent moths, and so will red cedar chips, or bits of cigar boxes.

To Tell Good Eggs.—Put them in water; if the butts turn up, they are not good.

To Avoid Kitchen Smells.—Put a few pieces of charcoal into the pots, kettles, and pans, when cooking.

To make Potatoes Mealy.—If the potatoes are watery, put a piece of lime, about as large as a hen's egg, into the pot, and boil with them, and they will become mealy.

A Superior mode of Boiling Potatoes.—Put them into a pot, with just sufficient water to cover them, leaving off the lid of the pot. When the water becomes scalding hot, without boiling, turn it off and replace it with cold water, adding salt. The cold water sends the heat from the surface to the heart of the potato, rendering it mealy. When potatoes are old, they should be pared previous to boiling them.

To Preserve Eggs.—Dissolve gum Arabic in water, making a thick solution. Coat your eggs with this, dipping them in one by one, and laying them out to dry. After they are dried, dip again the side on which they lay while drying, so as to coat them completely. When perfectly dried, pack them in powdered charcoal. An excellent process.

To Restore Wilted Flowers.—Put the stems one-third their length in scalding water, and let them stand until the water cools. Then cut off the scalded part of the stems, and put the flowers in cold water.

To Mend Cracks in Stoves, etc.—Mix wood ashes and common salt, equal parts, with water, to a thick paste. Fill the crack with this, whether the stove be hot or cold. It is a simple, cheap, but good cement.

To take off Starch or Rust From Flat-irons.—Tie up a

piece of yellow beeswax in a rag, and when the iron is almost, but not quite, hot enough to use, rub it quickly with the wax, and then with a coarse cloth.

INSECTS.—*To keep Red Ants from Closets.*—Throw some twigs of tomato vines on the shelves, or let the shelves be made of black walnut. Either will drive them away.

To Destroy Flies.—To one pint of milk add a quarter pound of raw sugar, and two ounces of ground pepper; simmer them together eight or ten minutes, and place the mixture about in shallow dishes. The flies attack it greedily, and are soon suffocated. By this method, kitchens, etc., may be kept clear of flies all summer without the danger attending poison. It is easily tried.

To Destroy Crickets.—Put Scotch snuff upon the holes where they come out.

TO PREVENT THE CREAKING OF A DOOR.—Rub a bit of soap on the hinges.

HOW TO SELECT FLOUR.—1. Look at its color; if it is white, with a slightly yellowish or straw colored tint, it is a good sign. If it is very white, with a bluish cast, or with black specks in it, the flour is not good. 2. Examine its adhesiveness; wet and knead a little of it between the fingers—if it works dry and elastic it is good, if it works soft and sticky it is poor. Flour made from spring wheat is likely to be sticky. 3. Throw a little lump of dry flour against a dry, smooth, perpendicular surface; if it adheres in a lump, the flour has life in it: if it falls like powder, it is bad. 4. Squeeze some of the flour in your hand; if it retains the shape given it by the pressure, that, too, is a good sign. Flour that will stand all these tests, it is safe to buy. These modes were given by old flour dealers, and we make no apology for printing them, as they pertain to a matter that concerns everybody: namely, the quality of that which is "the staff of life."

To CLEANSE THE INSIDE OF JARS.—There is frequently some trouble in cleaning the inside of jars that have had sweetmeats, or other articles put in them for keeping, and that, when empty, were wanted for other use. This can be done in a few minutes, without scraping or soaking, by filling up the jars with hot water (it need not be scalding hot), and then stirring in a teaspoonful or more of pearlash. Whatever of the former contents has remained sticking at the sides and bottom of the jar, will immediately be seen to disengage itself, and float loose through the water. Then empty the jar at once, and if any of the former odor remains about it, fill it again with warm water, and let it stand undisturbed a few hours, or till next day; then empty it again, and rinse it with cold water. Wash vials in the same manner. Also the inside of kettles, or any thing which you wish to purify or clear from grease expeditiously and completely. If you can not conveniently obtain pearlash, the same purpose may be answered nearly as well by filling the vessel with strong lye, poured off clear from wood ashes. For kegs, buckets, crocks, or other large vessels, lye may be always used.

FOWLS.—*To make Hens Lay.*—To every dozen of hens give one teaspoonful of cayenne pepper with their food every other day. They will produce nearly double the quantity of eggs.

To make Hens Lay all Winter.—Feed them freely upon sunflower seed, with other food, all winter.

Cure for Hen Cholera.—Boil the bark of a persimmon in an iron kettle. When the strength is abstracted, take out the bark and stir corn meal into the liquid till you form a thick dough. Give this dough to the hens. I have tried this remedy, and have no doubt it is reliable.

Dr. Trimble says, coop a hen with small chickens near your cucumber and melon vines, and the little chicks will take care of the striped bugs effectually, without injury to the vines—eat them up, and prevent them eating your vines. This is the most rational prescription for the evil that we have seen.

PICKLES AND PRESERVES.—*General Directions for Preserving Fruits.*—Great improvements have of late been made in the art of preserving fruits for family use, by the introduction of jars which can be hermetically sealed. The process of preserving is so simple, that every housekeeper can accomplish it—the only secret of success being that the fruit should be put up and sealed while hot, the jars being filled to the brim. The best jars for this purpose are those which are made entirely of glass. These will pay for themselves in a year or two, as fruit which is sealed so as to exclude the air may be preserved with one quarter the amount of sugar required in the old process, and retains its original flavor better.

The following directions for preserving in hermetically sealed jars will be interesting to housekeepers at the present time:

Select only good fresh fruit or vegetables. Stale and fermented articles can never be preserved, nor the decay already commenced arrested. Be particular and know to a certainty that your articles are fresh. No vegetables except tomatoes can be procured in the markets of large cities fresh enough for preserving.

Blackberries, Raspberries, and Strawberries.—Use from a quarter to a half pound of sugar to a pound of fruit. Sugar the strawberries, and let them stand for half an hour, then put the syrup which will be formed by the juice and sugar into a preserving kettle, and boil as long as any scum arises, and then put in the strawberries and boil until they are thoroughly heated through.

Fill the jars—after first warming them in some way—and close immediately, while the contents are hot.

Cherries and Blackberries.—Stew with or without sugar ten minutes, and seal up while boiling hot.

Gooseberries.—These can be kept by putting them into jars as they come from the bushes, and sealing them up. Wash and pick them when wanted.

Currants.—Heat to boiling point with sugar, and seal up boiling hot.

Plums.—Make a syrup, using about half a pound of sugar to a pound of fruit. Boil the plums in this syrup until the fruit is tender; then fill the jars, and close up while hot.

Peaches.—Pare and cut out the stones. Make a syrup, using from a quarter to half a pound of sugar to a pound of fruit. Boil the syrup five or ten minutes; then put in the peaches and boil until they are thoroughly heated through, and then fill the jars and close immediately.

Quinces.—Peel and quarter them, and boil in water until tender, then do them in the same way as peaches.

Pears.—Same as quinces, except that they require less sugar.

Apples.—Pare, quarter and boil until tender, but not long enough to break in pieces; then add as much sugar as will sweeten to the taste, and let the whole boil two or three minutes. While hot, pour into jars and close up.

Tomatoes.—Take off the skin, and boil them one hour, or cook them sufficiently for the table. Season to the taste, fill the jars, and close up boiling hot. These being a very juicy article, require much longer boiling than most other things to boil the water away.

If the above proportions of sugar make the fruit sweeter than is desirable, it can be kept with rather less, but green fruit requires more than ripe.

To Clarify Sugar for Preserving.—Put into a preserving pan as many pounds of sugar as you wish; to each pound of sugar put half a pint of water, and the white of an egg to every four pounds; stir it together until the sugar is dissolved, then set it over a gentle fire; stir it occasionally, and take off the scum as it rises; after a few boilings up, the sugar will rise so high as to run over the side of the pan; to prevent which, take it from the fire for a few minutes, when it will subside, and leave time for skimming, until a slight scum or foam only will rise, then take off the pan, lay a slightly wetted napkin over the basin, and then strain the sugar through it; put the skimmings into a basin; when the sugar is clarified, rinse the skimmer and basin with a glass of cold water, and put it to the scum, and set it by for common purposes.

To Preserve Gooseberries.—Take full-grown gooseberries before they are ripe, pick them, and put them into wide-mouthed bottles, cork them gently with new, soft corks, and put them in an oven, from which the bread has been drawn, and let them stand till they have shrunk nearly a quarter; then take them out and beat the corks in tight, cut them off level with the bottle, and rosin them down close. Keep them in a dry place.

Preserved Currants to eat with Meat.—Strip them from the stem. Boil them one hour. Add a pound of sugar to one pound of fruit. Boil all together twenty minutes.

Preserved Cherries.—Stone them. Allow one pound sugar to one pound fruit; put a layer of fruit at the bottom of the preserving kettle, then a layer of sugar, and repeat till all are in; boil till clear. Put in bottles hot, and seal them. Keep them in dry sand.

Preserved Pears.—Take out stems and cores, and pare them. Boil in water till tender; do not break them in taking out. Make a syrup of a pound of sugar to a pound of fruit, and boil the fruit in the syrup till clear.

Preserved Oranges.—Boil the oranges in water till you can run a straw through the skin. Clarify three-quarters of a pound of sugar for each pound of fruit. Take the oranges from the water and pour the hot syrup on them. Let them stand one night. Next day boil them in the syrup till it is thick and clear.

Purple Plums Preserved.—Take an equal weight of fruit and nice sugar, and fill a clean stone jar with the fruit and sugar in layers. Cover them, and set the jar in a kettle of water over the fire. Let them stand in the boiling water all day, filling up the kettle as the water boils away. If at any time they seem likely to ferment, repeat this process. It is a simple and excellent way of preserving plums.

Preserved Pumpkin.—Cut a thick yellow pumpkin, peeled, into strips two inches wide, and five or six long. Take one pound of sugar for each pound of pumpkin, and scatter it over the fruit, pouring on two wine-glasses lemon juice to each pound. Next day put the parings of two or three lemons in with the sugar and fruit, and boil the whole three-fourths of an hour, or long enough to make it tender and clear, without breaking. Lay the pumpkin to cool, strain the syrup, and pour over the pumpkin.

To Preserve Cucumbers.—Take firm ripe cucumbers, as soon as they turn yellow; pare them, take out the seeds, cut them in pieces two or three inches in length and about two in width; let them lie in weak salt and water for eight hours. Then prepare a syrup of one gallon of cider vinegar, five pounds of sugar, one ounce of mixed spices (not ground spices), boil twenty minutes, then strain. After drying the cucumber with a soft cloth, put it in the syrup, and boil till soft and transparent; skim the pieces out carefully, lay them in a colander to drain; then boil the syrup to the consistency of molasses, pour it on the cucumber, and keep in a cool place.

Fine Pickled Cabbage.—Shred red and white cabbage; spread it in layers in a stone jar, with salt over each layer. Put two spoons whole black pepper, and the same quantity each of allspice, cloves, and cinnamon, in a bag, and scald in two quarts of vinegar. Pour this vinegar over the cabbage, and cover it tight. It will be ready for use in two days.

Green Tomato Pickle.—One peck tomatoes, eight green peppers, to be chopped fine and soaked twenty-four hours in weak brine; then skim out, and add one head of cabbage chopped fine; scald in vinegar twenty minutes; skim it out and put in the jar, and add three pints of grated horse radish, and spices as you please. Pour cold vinegar over the whole.

Pickled Ripe Tomatoes.—To one gallon of peeled tomatoes, add two tablespoons white mustard seed, one tablespoon whole cloves, one tablespoon salt, two tablespoons pepper, two tablespoons allspice. Put in a jar, sprinkling the spices between the layers, and pour scalded vinegar over them.

Citron Melon.—Two lemons to one pound melon, equal weight of sugar for the fruit. Remove the pulp of the melon, cut it in thin slices, and boil in water till tender. Take it out, and boil the lemon in the same water twenty minutes. Take out the lemon, boil the sugar in the same water, adding a little more water, if necessary. When the syrup is clear, put in the melon and boil a few minutes.

Currant Jelly.—Four quarts ripe currants mashed in both hands, till nearly all are broken : squeeze out the stems and remove them. Put the pulp in a strong bag and squeeze very tightly, and there will be nearly three pints juice. Put three pounds of white sugar to this, and boil half an hour.

Currant Jelly without Cooking.—Press the juice from the currants, and strain it; to every pint put a pound of fine white sugar; mix them together until the sugar is dissolved; then put it in jars, seal them, and expose them to a hot sun for two or three days.

Tomato Figs.—Scald and remove the skin of eight pounds tomatoes; cook them in three pounds sugar, till they are clear; take them out with a spoon (with as little juice as possible), on dishes, to be dried in the sun, or a cool oven, occasionally turning them. When dry, pack them in a box, sprinkling sugar between the layers: these taste like figs. The round middle size tomatoes are the best.

Raspberry Jam.—Allow a pound of sugar to one pound fruit. Boil the fruit half an hour; strain one-quarter of the fruit, and throw away the seeds; add the sugar, and boil the whole ten minutes.

Black Butter.—Put to any kind of ripe berries half their weight of brown sugar; mash and stew them gently for half an hour, stirring them frequently. This is a good substitute for butter spread on bread, and is usually much liked by children, and is more healthy than butter, particularly for those afflicted with humors in the blood.

To Protect Dried Fruit from Worms.—It is said that dried fruit put away with a little sassafras bark—say a large handful to a bushel—will keep for years, unmolested by these troublesome little insects, which so often destroy hundreds of bushels in a season. The Remedy is cheap and simple.

To Keep Preserves.—Apply the white of an egg, with a suitable brush, to a single thickness of white tissue paper, with which cover the jars, overlapping the edges an inch or two. When dry, the whole will become as tight as a drum.

To Prevent Preserves Graining.—Add a teaspoonful of cream tartar to every gallon of the jam or preserves.

Citron Preserves.—Prepare the rind; boil very hard for half an hour, in tolerably strong alum water; then take them out and put them into clear cold water, and let them stand over night; in the morning change the water, and put them to boil, and let them boil until they have entirely changed their color, and are quite soft; then make the syrup, allowing one and a half pounds of white sugar to one pound of fruit; then add the fruit, and boil only a few minutes. Flavor with mace, cinnamon, ginger, or lemon, to taste. This makes a very good preserve.

Apple Preserves.—Peel, cut, and core the apples; and to

each pound of apples allow a pound of brown sugar. To every three pounds of sugar add one pint of water, and boil pretty thick, skimming it well; then add the apples and the grated peel of one or two lemons, and boil till the apples fall, and look clear and yellow. This is good, and will keep for years.

Peach Preserves.—Pare and cut the peaches, and put them into the preserve kettle, with sugar and peaches used alternately, taking about three-fourths of a pound of sugar to a pound of peaches, and let them remain so over night. By morning the sugar will be dissolved, and there will be enough of syrup formed to commence the cooking. Boil slowly until the peaches are thoroughly cooked.

Plum Preserves.—To a pound of plums allow a pound of sugar; add a little water, and boil slowly until the plums are cooked done; then take out the plums and put them into the jars, and boil the syrup down until it is as thick as desired, and pour it over the plums.

SEALING-WAX, OR CEMENT—*for Fruit Cans, Jars, etc.*—Melt together six ounces of rosin, four ounces of Venice turpentine, and two ounces of shellac. Color with lampblack, if desired.

Another.—Melt together one pound of rosin and one ounce of tallow.

WATER.—*To make Water Cool for Summer.*—The following is a simple mode of rendering water almost as cold as ice: Let the jar, pitcher, or vessel used for water be surrounded with one or more folds of coarse cotton, to be constantly wet. The evaporation of the water will carry off the heat from the inside, and reduce it to a freezing point. In India and other tropical regions where ice cannot be produced, this is common. Let every mechanic or laborer have at his place of employment two pitchers thus provided, and with lids or covers—the one to contain water for drinking, the other for evaporation—and he can always have a supply of cold water in warm weather. Any person can test this by dipping a finger in water, and holding it in the air of a warm day; after doing this three or four times, he will find his finger uncomfortably cold.

The Cheapest Filter.—Here is a method of constructing an excellent filter in the cheapest manner: Take a flower pot, or any other vase having a hole in the bottom, fill the bottom with large round pebbles, cover with smaller pebbles, then with coarse sand or fine gravel, and finally, with four inches of pounded charcoal. The charcoal may be placed in a bag, and broken with a mallet or hammer; it should then be sifted, and the very finest dust thrown away. A clean flannel, held down by stones on the corners, should cover the charcoal, which must be freely burned, and renewed occasionally.

Simple Mode of Purifying Water.—It is not generally known that pounded alum possesses the quality of purifying water. A tablespoonful of pulverized alum sprinkled into a hogshead of water (the water stirred at the same time), will, after a few hours, by precipitating to the bottom the impure particles, so purify it that it will be found to possess nearly all the freshness and clearness of the finest spring water. A pailful, containing four gallons, may be purified by a single teaspoonful of the alum.

To Make Hard Water Soft.—Add to one bucket of water, warmed, one ounce of carbonate of soda, which renders it soft as rain water.

Another Method.—A teaspoonful of salt thrown into the water will soften from three to four pails of hard water. This is a valuable receipt for housekeepers, and one which may be easily tested.

Another.—A half ounce quick lime dipped into nine quarts of water, and the clear solution put into a barrel of water—the whole will be soft as it settles.

To Soften Hard Water, or Purify River Water, simply boil it, and then leave it to atmospheric exposure.

To Save Ice from Melting.—The following method of preserving ice we publish for the benefit of such of our readers as are not able to procure ice boxes. It is a cheap, and, we expect, a first-rate contrivance:

Put the ice in a deep dish or jug, cover it with a plate, and place the vessel on a pillow stuffed with feathers, and cover the top with another pillow carefully, by this means excluding the external air. Feathers are well-known bad conductors of heat, and in consequence the ice is preserved from melting. Dr. Schwartz states that he has thus preserved six pounds of ice for eight days. The plan is simple, and within the reach of every household.

Curing Meat, etc.—*Our Receipt for Curing Meat.*—To one gallon of water allow one and one half pounds of salt, half pound of sugar, half ounce of saltpetre, half ounce of potash. In this ratio the pickle is to be increased to any quantity desired. Let these be boiled together until all the dirt from the sugar rises to the top, and is skimmed off. Then throw it into a tub to cool, and when cold, pour it over your beef or pork, to remain the usual time, say four or five weeks. The meat must be well covered with pickle, and should not be put down for at least two days after killing, during which time it should be slightly sprinkled with powdered saltpetre, which removes all the surface blood, etc., leaving the meat fresh and clean.

Some omit boiling the pickle, and find it to answer well; though the operation of boiling purifies the pickle by throwing off the dirt always to be found in salt and sugar.

If this receipt is properly tried, it will never be abandoned. There is none that surpasses it, if so good.

To preserve Hams.—Grind some black pepper fine, and put in a box; and as soon as the hams are well smoked, take them down, and dust the pepper over the raw part, and over the back; then hang them up in the smoke-house again.

To Keep Dried Beef or Hams from mold, bugs, and every form of decay. After curing and smoking, hang in a cool, dry place, and once in two or three weeks rub the meaty part thoroughly with cider brandy, high wines, or alcohol. I will warrant it, thus prepared, to keep as good as new.

Receipt for Curing Hams.—Six gallons soft or rain water, nine pounds rock salt, three ounces saltpetre, one and a half ounces pearlash, one quart molasses, three pounds brown sugar. The saltpetre and pearlash must be dissolved before putting into the pickle. Boil and skim. Rub the hams thoroughly with fine salt, before putting them into the pickle, to get the blood out of them.

To Pack Pork.—Scald coarse salt in water, and skim it till the salt will no longer melt in the water; pack your pork down in light layers; salt every layer; when the brine is cool, cover the pork with it, and keep a heavy stone on the top, to keep the pork under brine. Look to it once in a while for the first few weeks, and if the salt has all melted, throw in more. This brine, scalded each time used, will keep good twenty years.

How to Pack Pork for Family Use.—In the first place, have a good hog, dress him in a cool time, and let him hang until the whole carcass is cooled throughout. Cut out the fat pork neatly, taking every piece that is bloody, or wash out the blood clean with cold water. The strips should be three or four inches wide. Lay them in a cool place, each piece by itself, for twenty-four hours; then sprinkle the bottom of a perfectly sweet barrel, half an inch thick, with what is called "coarse food" salt, and on that lay the strips of pork, with the skin next the barrel, and so continue until the bottom is entirely covered. Pack it closely; then cover that layer with half an inch of salt, and so continue until the barrel is nearly full; cover the top with salt, and then lay a clean strip of board over it, and on that a stone sufficiently heavy to keep the whole from rising. Let it stand twelve hours, and then fill the barrel with clean, cold water.

Best Method of Keeping Beef.—Cut up the meat in pieces as large as you desire. Pack it in a barrel or cask. Then make a brine as follows: one and one half pounds salt to one gallon of water, one ounce saltpetre to one hundred pounds of beef, one tablespoonful of ground pepper to one hundred pounds of beef. Put in salt and saltpetre, and heat it boiling hot, skim it, and then add the pepper; put it on the beef, boiling hot, and cover closely. Your meat will be good at any time. The philosophy is this: the hot brine closes the pores on the surface, preventing decay, and the meat from getting too salt. Try it. If necessary, scald the brine over in the spring, or put on a new brine. Farmers can in this way have fresh meat nearly all the time. The meat should be taken as soon as it gets cold, before it has acquired any old taste by exposure to the atmosphere.

Making Sausages.—To twenty pounds of meat cut fine, add the following: powdered sage, one ounce; powdered summer savory, one ounce; salt, five ounces; black pepper, two ounces. Work the mass well together, and put in cases or bags, as taste may direct.

To Keep Sausage Meat.—Prepare it in small round cakes, fry them as for the table, pack them closely in an earthen jar, pour the fat from the frying pan over them, and put a weight on them to keep them down until cold, then remove the weight and cover the top with lard. Keep dry and cool.

CIDER.—*To Preserve Cider.*—When the cider in the barrel is in a lively fermentation, add as much white sugar as will be equal to a quarter or three-quarters of a pound to each gallon of cider (according as the apples are sweet or sour), let the fermentation proceed until the liquid has the taste to suit, then add a quarter of an ounce of sulphite of lime to each gallon of cider, shake well, and let it stand three days, and bottle for use.

The sulphite should first be dissolved in a quart or so of the cider before introducing it into the barrel of cider.

Sour Cider.—A gill of mustard seed to a barrel of sour cider will return it to its sweetness, or prevent its turning sour, if still in good order. A quarter of a pound of saltpetre to a barrel of cider will also prevent it from change.

To Keep Cider Sweet.—Put to a barrel of new cider a gill of white mustard seed. This will prevent it from becoming hard and sour for many months. If you wish to keep it from fermenting, put into the barrel a bag containing pulverized charcoal. Treated in this way, it will not possess any intoxicating qualities, and improves by age. In bottling cider, put into each bottle three or four raisins, to make it brisk.

To Make Cider without Apples.—Soft water, four gallons; brown sugar, three pounds; tartaric acid, two ounces; cream tartar, two ounces. Mix, and when dissolved, it is ready for use.

WINES.—*To make Cider Wine.*—Let your cider ferment, then heat it till it boils. Skim it, and add to each gallon of cider one pound of sugar, and one pint of whisky. To give it a high color, boil in the cider a small bag of dried black raspberries.

To Make Champagne from Apples.—Press the juice directly from sound apples without grinding; let the juice run directly from the press into a filter consisting of a box about twelve inches deep by six inches square, filled with a mixture of pulverized charcoal and clean sand or fine gravel in about equal quantities; the bottom of the box is perforated with small holes, and covered with clean straw before the filtering material is put in. The juice is passed through this filter into bottles, which are immediately corked. We have tasted cider a year old, made and put up in this manner that had much of the flavor of genuine champagne wine, and in every respect a pleasant beverage.

Cherry Wine.—Water, five gallons; cherries, five gallons; sugar, fifteen pounds; red tartar, one ounce; brandy, two pints. This will make nine gallons. If garlic is added, it will kill worms in children, and is a first rate tonic at all times. Use wild cherries, if good and handy.

Grape Wine.—Wine is a thing not made by man at all, but only modified, at most. It is a production of nature. In the purest and best grape wine, this is most remarkably exhibited. The grapes are easily pressed by a wine, or even cider press, and can be kept separate from the lees or allowed to ferment on them, as strength is required. Not one drop of water, not even a lump of sugar is requisite, though most of the wines in this country are made with it. But the fermentation is all an act of nature herself. She it is who makes our wine, and all that men have to do while fermentation is going on in the juice, is to watch it and let it alone. And when this has ceased, the drawing off into a clean cask, and keeping it undisturbed in quiet and perfect darkness by itself, is all that man can do. A lump of loaf sugar in each bottle may give a champagne freshness to it, but the simpler, the purer, the less of cookery in wine, the better for it and those who have good taste enough to prefer it thus. The pure juice of grapes is best in sickness. The best of grapes, and, if sugar be added, the best and purest sugar, should alone be used.

How to make Blackberry Wine.—There is no wine equal to the blackberry wine, when properly made, and all persons who can conveniently do so, should manufacture enough for their own use every year, as it is invaluable in sickness as a tonic, and nothing is a better remedy for bowel diseases. The following is the recipe for making it:

Measure your berries and bruise them, to every gallon adding one quart of boiling water. Let the mixture stand twenty-four hours, stirring occasionally; then strain off the liquor into a cask, to every gallon adding two pounds of sugar; cork tight, and let it stand until the following October, and you will have wine, ready for use, that will make the lips smack as they never smacked under similar influence before.

The following receipt for making this wine, is from Mrs. Hale's new cook book:

To five gallons of ripe blackberries, add seven pounds of honey and six gallons of water; boil, strain and leave the liquor to ferment; then boil again, and put into a cask to ferment.

The first recipe, it will be perceived, says nothing about *fermentation*, which is very important to be attended to, and which, unless it is, may burst the cask and spill the wine.

BLACKBERRY CORDIAL.—Two quarts of blackberry juice; one pound of loaf sugar; a quarter of an ounce of cloves; half an ounce of nutmeg; half an ounce of allspice; boil it all fifteen minutes. When cold, add a pint of brandy.

RUM, GIN, AND BRANDY.—*To make Artificial Gin.*—Pure Cologne spirits, four gallons; Holland gin, one gallon; oil of juniper, three drachms; oil of anise, one drachm. Mix well.

To make Artificial Rum.—Pure Cologne spirits, four gallons; good Jamaica or St. Croix rum, one gallon; oil of caraway, half a drachm. Mix well.

To make Artificial Brandy.—Pure Cologne spirits, two gallons; best French brandy, half a gallon; loaf sugar, quarter of a pound; sweet spirits of nitre, one ounce. Mix well, and color with burnt sugar.

These liquors are pure, and much better than those usually bought for such.

BEER.—*Spruce Beer.*—Mix thoroughly in a pail three

quarts of molasses, and one ounce of the essence of double spruce; to this may be added one pound of best ginger; fill the pail with boiling water; pour the mixture into a clean half barrel; fill it up with cold water; add a quart of yeast, and shake the whole well together; after fermenting one or two days, the bung may be put in, and it will be fit for bottling or for use.

Hop Beer.—Boil one handful of hops in one quart of water; strain it; add one teaspoon ginger, one pint of molasses, one pailful of lukewarm water, one penny's worth of yeast. Let it stand twenty-four hours; take off the scum and bottle it for use.

Cream Beer.—Two and one-fourth pounds white sugar, two ounces tartaric acid, juice of half a lemon, and three pints of water—boil together five minutes. When nearly cold, add the whites of three eggs beaten to a froth: one half cup of flour, well beaten, and half an ounce of wintergreen essence. Bottle, and keep in a cool place. Two tablespoonsful of this syrup in a tumbler of water, with one-fourth teaspoon of soda. It is ready for use as soon as made, but age improves it, and it will keep any length of time. Shake the bottle well every time before using.

VINEGAR.—To eight gallons of soft water add one gallon of alcohol, one quart of molasses, and a dozen white beans, done up in a brown paper, to form the mother. Let it stand two or three weeks, in a warm place.

Cider Vinegar.—There are hundreds of farmers in the western country, who are most of the time either destitute of vinegar entirely, or make use of some slops, which is not only unhealthy, but decidedly unpalatable. The vinegar manufactured from acids enters largely into the consumption of towns and cities, and to some extent into that of the country also. Whisky, with all its adulterations, is used for the purpose of making pickles, and in that manner lends its aid to the destroyer of human life. Many other different methods of procuring the sours of life are practised, and many of which are not only productive of deleterious influences to the health of ourselves and our children, but require far more labor than ought to be bestowed upon that branch of a housewife's business.

We live in an age of labor-saving machines, and we ought to economize, both in labor and money, as well in the less important matters of living, as in the most important. And to apply a little Yankee ingenuity in this case, is not so difficult as many people imagine. Almost every family in the country have the materials for manufacturing pure cider vinegar, if they will only use them. Common dried apples, with a little molasses and brown paper, are all you need to make the best kind of cider vinegar, if they will only use them. Common dried apples, with a little molasses and brown paper, are all you need to make the best kind of cider vinegar. And what is still better, the cider which you extract from the apples, does not detract from the value of the apples for any other purpose.

Soak your apples a few hours—washing and rubbing them occasionally, then take them out of the water, and thoroughly strain the latter through a tight woven cloth; put it into a jug, and add a pint of molasses to a gallon of liquor, and a piece of common brown paper, and set in the sun, or by the fire, and in a few days your vinegar will be fit for use. Have two jugs, and use out of one while the other is working. No family need be destitute of good vinegar, if they will follow the above directions.

MAKING SOAP, WASHING, ETC.—*One Hundred Pounds of Good Soap for* $1.30.—Take six pounds of potash, seventy-five cents; four pounds of lard, fifty cents; a quarter of a pound of rosin, five cents. Beat up the rosin, mix all together well, and set aside for five days; then put the whole into a ten gallon cask of warm water, and stir twice a day for ten days, at the expiration of which time you will have one hundred pounds of excellent soap.

To make Soap Without Boiling.—Allow one gallon of lye, strong enough to bear up an egg, to every pound of grease. Put the lye into your barrel, and strain the grease hot through a seive, or colander. Stir this three or four times a day, until it thickens. By this process, you have soap clean, and with much less trouble than in the old way.

Recipe for Making Soap.—Pour four gallons of boiling water over six pounds of washing soda and three pounds of unslaked lime; stir the mixture well, and let it settle until it is perfectly clear. It is better to let it stand all night, as it takes some time for the sediment to settle. When clear, drain the water off, put six pounds of grease with it, and boil for two hours, stirring it most of the time. If it does not seem thin enough, put another bucket of water on the grounds, stir and drain off, and add as is wanted to the boiling mixture.

Its thickness can be tried by putting a little on a plate to cool occasionally. Stir in a handful of salt just before taking it off the fire. Have a tub ready soaked—to prevent the soap from sticking—pour it in, and let it stand till solid, when you will have about forty pounds of white soap, at a cost of about two cents per pound.

To Whiten Linens.—Stains occasioned by fruit, iron-rust, and other similar causes, may be removed by applying to the parts injured a weak solution of the chloride of lime—the cloth having been previously well washed—or of soda, oxalic acid, or salts of lemon, in warm water. The parts subjected to this operation should be subsequently well rinsed in soft, clear, warm water, without soap, and be immediately dried in the sun.

To take out Mildew.—Mix soft soap with starch powdered, half as much salt, and the juice of a lemon; lay it on the part on both sides with a painter's brush. Let it lie on the grass day and night till the stain comes out.

For Making Soap.—Hard soap, three pounds; rain water, four quarts; sal soda, a half pound. Or, soft soap, two gallons; rain water, six quarts; sal soda, one pound.

For Making a Harder Soap.—Hard soap, one pound; rain water, one gallon; sal soda, half a pound; super-carbonate soda, two ounces; a small quantity of salt, about a tablespoonful to be added while boiling. All of the above are to be boiled until they assume a proper consistency and color. By adding to either of the above, while boiling, a little lime-water, it will improve them.

Directions for Making the Mixture for Washing.—To five gallons of soft water add half a gallon of lime-water, one pint and a half of soft soap, or half a pound of hard soap, and two ounces or two tablespoonsful of sal soda.

Method of Washing with the Mixture.—Soak the clothes over night, if very dirty; at any rate, wet them thoroughly before putting them into the mixture; when the mixture is at boiling heat, put in the clothes that have been soaked or wet, merely rubbing such parts with a little soap as are usually soiled; boil them one hour, then take them out and drain them, rinse them thoroughly in warm water, then rinse them in indigo water, as usual, and they are fit for drying.

Washing.—To save your linen and your labor: pour on half a pound of soda two quarts of boiling water, in an earthenware pan; put half a pound of soap, shred fine, into a saucepan with two quarts of cold water, stand it on a fire till it boils, and when perfectly dissolved and boiling, add it to the former; mix it well, and let it stand till cold, when it has the appearance of a strong jelly. Let your linen be soaked in water, the seams and any other dirty part rubbed in the usual way, and remain till the following morning. Get your copper ready, and add to the water about a pint basin full; when lukewarm, put in your linen, and allow it to boil twenty minutes. Rinse in the usual way, and that is all which is necessary to get it clean, and to keep it in good color. The above recipe is invaluable to housekeepers. If you have not tried it, do so without delay.

Superior Preparation for Starching.—Put into a pitcher a couple of ounces of gum Arabic, pour on a pint of boiling water, cover it over and let it remain until the succeeding day, then turn it off carefully from the dregs into a clean bottle, and cork it up for use. A tablespoonful of this, stirred into a pint of Poland starch, made in the usual manner, will give a fine gloss to linen, and will impart a look of newness to either white or colored lawns.

Starch Polish for Shirt Bosoms.—White wax, one ounce; spermaceti, two ounces. Melt them together with a gentle heat. A piece about the size of a large pea, put into a sufficient amount of starch for a dozen pieces, is all that is required.

Washing.—Except woolens and colored clothes, all other kinds should be put to soak over night, the very dirty parts having soap rubbed on them. If you use a washing fluid, it is usually mixed in the soaking water; if you use no wash mixture, the next morning wring out the clothes, and proceed to wash them carefully through two warm waters; then boil them in clean water rather briskly, but not longer than half an hour. Wash them out of the boil, rinse through two waters. The last rinse water should have a delicate tinge of blue, likewise a small quantity of starch for all cottons or linens; reserve those you wish stiffer for the last, and mix more starch in the water. Shirt bosoms and collars, skirts—in short, anything you wish very stiff, should be dipped while dry. Swiss and other thin muslins and laces are dipped in starch while dry, and then clapped with the hands until in a right condition to iron.

Calicoes, brilliants, and lawns of white grounds, are washed like any other white material, omitting boiling, until the yellow tinge they acquire makes it absolutely necessary. Unbleached cottons and linens follow the white clothes through the same waters, but must in no case be boiled with them, as they continually discharge a portion of their color, and so discolor the white clothes.

In directing the preparation for washing fluids, we give the process employed with them; but colored clothes, in our experience, can be washed in none of them without injury to the color.

Calicoes, colored lawns, and colored cottons, and linens, generally, are washed through two suds and two rinsing waters, starch being used in the last, as all clothes look better, and keep clean longer, if a little stiffened.

Many calicoes will spot if soap is rubbed on them; they should be washed in a lather, simply. A spoonful of ox gall to a gallon of water will set the colors of almost any goods soaked in it previous to washing. A teacup of lye in a bucket of water will improve the color of black goods.

Nankeen should lay in lye awhile before being washed; the lye sets the color.

A strong clean tea of common hay will preserve the color of those French linens so much used in summer by both sexes.

Vinegar in the rinsing water, for pink or green calicoes, will brighten them. Pearlash answers the same end for purples and blue.

Flannels should be washed through two suds and one rinsing water; each water should be as hot as the hand can bear, unless you wish to thicken the flannel. Flannels washed in lukewarm water, will soon become like fulled cloth.

Colored and white flannels must be washed separately; and by no means washed after cotton or linen, as the lint from these goods adheres to the flannel.

There should be a little blue in the rinsing water for white flannel. Allow your flannels to freeze after washing in winter—it bleaches them.

Prints with Fast Colors.—These should be washed in warm suds, and scalded, if the ground is white, by pouring boiling water over them, but if dark, the scalding should be omitted. Rinse thoroughly, and add a little starch to the last water, merely sufficient to give them a fresh look, without stiffening them. Iron as soon as possible after being starched. Prints soon sour.

To Wash Calicoes.—Soap and cold water will remove dirt and grease from calicoes, and they will retain their color to the last.

To Wash Flannels.—Put them into a tub, and pour cold suds on them, and let them soak for twelve hours; then wash with the same suds, and rinse in cold water. They will keep their color and not shrink.

FARMERS' DEPARTMENT.

DISEASES OF HORSES.

For Colic.—Mix half a pint of salt, half a pint of soft soap, and half a pint of soot, in half a gallon of warm water. Drench the horse with it, and give him exercise.

For Botts.—Drench the horse with half a gallon of new milk, sweetened with molasses. This usually will relieve the horse from the gnawing of the botts, which may be known by his becoming easy. In an hour or two after he becomes quiet, give him a strong dose of glaubers salts.

Another Remedy—(and in an advanced stage, more certain than the above).—Take half a gallon of blood from the neck of the sick horse (or some other one), and while it is warm, drench the horse with it. This will relieve the horse in a few minutes, as the botts will stop gnawing the horse, and fill themselves with this blood. In half an hour after the horse has become easy, drench him with a pint of whisky, in which human excrement has been mixed; and in an hour or two thereafter, give him a large dose of glaubers salts, or some other physic.

Another Remedy.—Beeswax, mutton tallow, and loaf sugar, each eight ounces; put them into one quart of new milk, and warm it until all is melted. Then put it in a bottle, and give it just before the wax begins to harden. About two hours afterwards, give a good dose of salts, or some other physic. The botts will be discharged.

STIFLE.—The following recipe for curing a stifle is considered invaluable, if not positively infallible. At any rate, many horses have been cured by it, and the recipe been sold for many dollars:

A handful of sumach bark, and a handful of white oak bark, boiled in a gallon of water down to two quarts; bathe the stifle with this lotion twice a day for three days; then put on a salve made of the white of an egg and rosin, and bathe the same in with a hot shovel two or three times, and the horse is cured.

WOUNDS IN HORSES.—One who claims to know something about horses commends the following remedy for the healing of wounds upon them;

Saltpetre should be moderately strong to taste, and bluestone added until the solution is slightly tinged. This, and nothing else, is to be used as a wash, two or three times a day. It purifies the wound, destroys the proud flesh, produces granulation immediately, and heals in a surprisingly short time. I have had a horse badly kicked and otherwise hurt, in midwinter, and midsummer, and their cure was equally rapid, and afterwards no scar was visible. The wound requires no covering (flies will not approach it) and dressing it with a mop of rags tied to a stick is very little trouble. Wounds do not need to be sewed up under this treatment—at least I never saw any advantage from it, as the stitches have uniformly torn out.

SCRATCHES.—Mix white lead and linseed oil in such proportion as will render the application convenient. But two or three applications are usually necessary to effect a common cure.

SPEEDY CURE FOR A FOUNDERED HORSE.—As soon as you find that your horse is foundered, bleed him in the neck in proportion to the founder—in extreme cases you may bleed him as long as he can stand up. Then draw his head up, as is common in drenching, and with a spoon put back on his tongue strong salt, until you get him to swallow one pint. Then anoint around the hoofs with turpentine, and your horse will be well in one hour.

A founder pervades every part of the system of a horse. The fleam arrests it from the blood, the salt arrests it from his stomach and bowels, and the spirits of turpentine arrests it from the feet and limbs.

Another Cure.—If a horse is foundered over night, he may be cured in three hours, if it is attended to in the morning. Heat a pint of hog's lard boiling hot, and after cleaning his hoof well, and taking off his shoe, put his foot in the lard, and with a spoon apply it to all parts of the hoof, as near the hair as possible. The application should be to the foot of each foundered limb.

Another Cure.—Dr. Thornton, of Virginia, a great breeder of horses, says the founder may be cured, and the horse fit for service the next day, by giving him a tablespoonful of alum. This is certainly an easy and cheap method of getting rid of that troublesome malady.

GLANDERS.—The following paragraph occurs in Dr. Dadd's new book on the horse:

"Whoever undertakes to attempt the cure of this awful malady must remember that he is running a great risk of losing his own life, for the absorption of the least particle of the virus will cause death in one of the most horrible of all forms; and many cases are on record going to show that whole families have been destroyed by absorbing the glandered virus."

COUGH IN HORSES.—The boughs of the common cedar,

cut fine, and mixed with the food of horses, are said to be an effectual remedy for the troublesome and very prevalent disease called "cough."

FOR POLL EVIL, AND FISTULA.—After they begin to discharge their matter, take salt and soft soap, in equal parts, mix and simmer dry, stirring, so as to make a powder, and insert this freely, and it will cleanse them out and cause them to heal.

PHYSIC BALL FOR HORSES.—Cape aloes, from six to ten drachms; Castile soap, one drachm; spirits of wine, one drachm; syrup to form the ball. If mercurial physic be wanted, add from one half a drachm to one drachm of calomel.

Previous to physicking a horse, and during its operation, he should be fed on bran mashes, allowed plenty of chilled water, and have exercise. Physic is always useful; it is necessary to be administered in almost every disease. It improves digestion, and gives strength to the lacteals, by cleansing the intestines and unloading the liver; and if the animal is afterwards properly fed, will improve his strength and condition in a remarkable degree. Physic, except in urgent cases, should be given in the morning, and on an empty stomach; and if required to be repeated, a week should intervene between each dose.

Before giving a horse a ball, see that it is not too hard or too large. Cattle medicine is always given as a drench.

TONIC FOR HORSES AND CATTLE.—Sulphate of copper, one ounce to twelve drachms; white sugar, one half ounce. Mix, and divide into eight powders, and give one or two daily in the animal's food.

COUGH BALL FOR HORSES.—Pulverized ipecac, three-fourths ounce; camphor, two ounces; squills, one half ounce. Mix with honey to form into mass, and divide into eight balls. Give one every morning.

Fever Ball.—Emetic tartar and camphor, each, one half ounce; nitre, two ounces. Mix with linseed meal and molasses to make eight balls. Give one twice a day.

Worm Ball.—Assafœdita, four ounces; gentian, two ounces; strong mercurial ointment, one ounce. Make into mass with honey. Divide into sixteen balls. Give one or more every morning.

Purgative Ball.—Aloes, one ounce; cream tartar and Castile soap, one-fourth ounce. Mix with molasses to make a ball.

Cathartic Powder.—To cleanse out horses in the spring, making them sleek and healthy. Black sulphuret of antimony, nitre, and sulphur, each equal parts. Mix well together, and give a tablespoonful every morning.

Blistering Liniment.—Powdered Spanish flies, one ounce; spirits turpentine, six ounces. Rub on the belly for pain in the bowels, or on the surface for internal inflammation.

Liniment for Sprains, Swellings, etc.—Aqua ammonia, spirits camphor, each two ounces; oil origanum and laudanum, each one half ounce. Mix.

Lotion for Mange.—Boil two ounces tobacco in one quart water; strain; add sulphur and soft soap, each two ounces.

Uterine Stimulant.—Infuse a handful of rue or savin in two quarts of water, and add two ounces salt.

To Keep Flies From Horses.—Take a handful of walnut leaves, upon which pour a quart of cold water; let it infuse one night, and pour the whole, next morning, into a kettle, and let it boil for a quarter of an hour; when cold it will be fit for use. Moisten a sponge, and before the horse leaves the stable, let those parts which are most irritable be washed with the liquor: viz., between and upon the ears, the neck, the flank, etc. Not only the lady or gentleman who rides out for pleasure will derive benefit from this preparation, but the coachman, the wagoner, and all others who use horses during the summer months.

Flies will not alight, for a moment on any thing bathed with pennyroyal. This simple remedy ought to be in every livery stable and country inn. It would save horses and cows a great deal of suffering.

Another Remedy.—Take one part coal oil, and two parts lard. Mix and rub over the ears and neck.

To Start a Baulky Horse.—Fill his mouth with dirt or gravel from the road, and he'll go. Now, don't laugh at this, but try it. The plain philosophy of the thing is, it gives him something else to think of.

For Cattle.—*To Relieve Choked Animals.*—Take a flexible rod about four feet long, and three-fourths of an inch in diameter, wind on the butt end tow or cotton, and tie a rag over that, and grease it. To keep the mouth open, place a piece of hard wood one foot long, four inches wide, and one inch thick, with a hole bored in the center for the rod to pass through, and then push it gently down the throat, and it is said to be always effectual, and to give immediate relief. It is also said that a round stick about the size and length of a rolling-pin, with a cord tied in the notches in the ends, placed in the animal's mouth, and fastened to each horn, will, if allowed a little time, unchoke them, and save the suffering creature from a lingering death.

If they are choked with apples, potatoes, turnips, or the like, and which can be felt in the throat from the outside, hold a solid block of wood against the article, on one side of the neck, and with a mallet strike upon the article from the other side.

Physic for Cattle.—Cape aloes, four drachms to one ounce; epsom salts, four to six ounces; powdered ginger, three drachms. Mix and give in a quart of gruel. For calves, one third of this will be a dose.

Pimento (allspice) Tea has been proved a sovereign remedy for diarrhœa in calves. Two tablespoonsful of ground pimento, put into three gills of boiling water, is sufficient for a potion, and should be given once in twelve hours, till relieved.

Miscellaneous.—*Experiment in Germinating Corn.*—Four boxes of earth, alike in quality and exposure to light and heat, were planted at the same time with corn from a single ear, and placed recently in a physician's office. In one box dry corn was planted; in another, the seed was soaked in clean warm water; in the third, the seed was soaked in a solution of lime water; in the fourth, the seed was soaked in chloride of lime and copperas water, equal parts. One week afterwards, the box planted with dry corn had not germinated, the second box had just commenced to germinate, the third box was just showing its green blades, and in the fourth box the blades were nearly three inches high. Copperas will keep birds and worms from eating the seed, and one pound will soak seed enough for twenty acres.

To Increase the Weight of Grains, etc.—Fill an open-mouthed vessel with water, and sink it partly down into the heap of grain. Soon the dry grain will absorb the water, and thus measure and weigh more. Moisten silk, and it will weigh more. Add ashes of bones to soap: bean meal to wax; chestnut meal to honey; boil rosin in oil, etc. This is all roguery and swindling, but I give it as information only.

To Raise Double Crops, Etc.—Throw a solution of sulphur and salt on your dung, before you spread and plow it in. The same will cause double crops of grass, and, in fact, of every grain and vegetable that is raised. It is a hundred times better than plaster and guano, mixed.

To Raise Grass, Clover, Mushrooms, Etc., without Seed.— Spread a little lime on waste moss ground, and you will get an abundant crop of clover. Cow and horse manure mixed will produce mushrooms. Oats sown at the usual time, and kept eaten down or cropped down without getting ripe, will, the next season, from the same stalks, produce an abundant crop of rye. I can only account for these things upon the simple ground that the most primitive types under a law to which that like production is subordinate, gives birth to the type next above it; that this again produced the next higher, and so on, to the very highest known existence.

Potato Rot Preventive.—A correspondent informs the "New York Sun," that after drying his potatoes a few hours in the field, he placed them in barrels, distributing in each barrel half a peck of quick lime, well mixed, with an equal quantity of powdered charcoal, which kept the potatoes sound all the year. He thinks the same mixture would prevent rot in the hill, if spread on the manure; but this he has not tried.

Chinch Bug.—A correspondent of the Agricultural Bureau at Washington gives his method of destroying the chinch bug last year. He says:

"On the 14th of June I sowed one bushel of salt on one

acre, and in thirty hours afterwards they had all left the piece of wheat, and gone northeast, into the corn. That one acre of wheat was all I got from the ten, and from it I threshed seventeen bushels."

AN INSECT TRAP.—Scoop out the inside of a turnip, scollop the edges, and place it downward on the earth. The insects will pass into it as a place of retreat through the holes; and the cucumbers, squashes, melons, etc., may soon be clear of them.

TO PROTECT GRAIN, FRUIT, AND PLANTS FROM INSECTS.—The leaves of the elder, if strewn among corn, or any other grain, when it is put in the bin, will effectually preserve it from the ravages of the weevil. The juice will also kill bedbugs and maggots. Insects never touch elder bushes. The leaves of the elder scattered over cabbages, squashes, cucumbers, and other plants subject to the ravages of insects, effectually shields them. The plum and other fruits may be saved by placing on the branches and among them, bunches of elder leaves.

GRAFTING.—Melt beeswax and tallow together, stirring in a little chalk while hot; dip in some strips of rag; then tear them into strips suitable to envelop the stock and scion; let the stock and scion be so covered as to prevent the escape of the sap, or the introduction of water, and the work is finished.

PEACHES WITHOUT STONES.—To make peaches grow without stones, an agriculturalist, who has tried it with success, says:

"Turn the top of the tree down, cut off the ends, stick them into the ground, and fasten so with stakes; in a year or so these tops will take root, and when well rooted, cut the branches connecting these reversed and rooted branches with the tree proper, and this reversed peach tree will produce fine peaches without stones." The same experiment may be tried with plums, cherries, and currants.

Trees Girdled.—These may be saved by taking, in the spring of the year, a piece of bark off from any limb, and putting it on where the bark is gnawed off—using care to fit it nicely at the ends—and then covering it all over with grafting clay. It is not material that the bark so put on should extend entirely around the tree; if a channel is formed for the flow of the sap, the tree will be saved.

At the last meeting of the Indiana Horticultural Society, it was stated that the best way to prevent rabbits from girdling trees, was to smear the bark of the trees with blood. If the trees are rubbed with a piece of fresh liver, the rabbit will not touch the bark as long as the least taint of blood remained, even after being thoroughly washed by rains.

To Protect Fruit Trees from Curculio.—Saturate sawdust with coal oil, and place it around the roots of the trees.

Peach Worm.—Plant tansey around the roots of peach trees. The peach worm will not trouble them afterwards.

A Hartford plumber has accidentally discovered that the smoke from a little charcoal fire under a tree will suffocate hundreds of worms upon it. A little sulphur placed on hot embers answers the same purpose.

To Bring Dead Trees to Life.—Bore a deep hole near the roots, and fill it nearly full of blue vitriol. If there is any life remaining in the roots, it will soon be reinvigorated, and flourish with exceeding beauty. It is by this process that different substances may be made to produce the fruit of all trees, vines, bushes, and even vegetables, of the kinds that grow on top of the ground.

Cure for Hog Cholera.—For every twelve hogs take one gallon of soft soap, half a gallon common grease, and five cents' worth of salæratus. Let it be warmed, and added to a slop of wheat bran and water, and fed while warm.

Itch in Swine.—Rub the animals with equal parts of lard and brimstone, and put them in a clean pen.

How to Get Rid of Rats.—For some years I was considerably annoyed with rats. I tried various vermin poisons, traps, etc., with very little success, until I thought of a mode which we adopted for destroying dogs that used to hunt our rabbit warren in the old country. So I had a quantity of broken bottles and window glass, and with a hammer and an anvil, triturated it pretty fine (a stone would do to pound the glass on); I then sifted the coarse part out, and mixed a cupful of the fine with a cupful of flour and another of oatmeal, and scenting it with a few drops of aniseed to attract them, I placed it on boards in the cellar, etc. They ate it up so fast that one of the family observed that instead of poisoning, it must be fattening them; but a few days told a different story. The last mess served for them remains untouched yet, though put down last fall, and no appearance of rat or mouse, living or dead, since. Neither have we noticed any smell, or blue-bottle (meat) flies, as there would have been had they died on the premises. It was a happy riddance. The mixture must be kept from children, dogs, and other silly animals, as it would kill them as well as rats.

Facts in Cutting Timber.—Cut timber from the middle of September to the middle of December, and you cannot get a worm into it. October and November are perhaps the best months, and sure to avoid the worm.

You cut from March to June, and you cannot save the timber from worms or borers. May used to be called peeling time; much was then done in procuring bark for the tanneries, when the sap is up in the trunk, and all the pores full of sap; whereas in October these pores are all empty—then is the time to cut, and there will be no worms.

When you see an ox-bow with the bark tight, there are no worms, no powder-post, and you can not separate it from the wood; and what is true in one kind is true in all kinds of timber, and every kind has its peculiar kind of worm. The pine has, I believe, the largest worms; these worms work for many years. I have found them alive and at work in white oak spokes that I knew had been in my garret over twelve years, and they were much larger than the first; they do not stop in the sap, but continue in the solid part. I do not think of buying timber unless it is cut in the time above alluded to.

I have wondered that there has not been more said on this subject, as it is one of great importance, even for fire-wood, and especially for ship building, etc.

To Keep Tires on Wheels.—Hear what a practical man says of this subject:

"I ironed a wagon some years ago for my own use, and before putting on the tires I filled the felloes with linseed oil; and the tires have worn out and were never loose. I also ironed a buggy for my own use seven years ago, and the tires are as tight as when put on. My method of filling the felloes with oil is as follows: I use a long cast iron heater, made for the purpose. The oil is brought to a boiling heat, the wheel is placed on a stick so as to hang in the oil, each felloe an hour for a common sized felloe. The timber should be dry, as green timber will not take oil. Care should be taken that the oil does not get hotter than the boiling heat, in order that the timbers are not burnt. Timber filled with oil is not susceptible to water, and is much more durable. I was amused, some years ago, when I told a blacksmith how to keep the tires tight on wheels, by his telling me that it was a profitable business to tighten tires; and the wagon maker will also say that it is profitable to him to repair wheels. But what will the farmer, who supports the wheelwright and smith, say?"

MEASUREMENT OF CORN IN THE CRIB.—After leveling the corn, multiply the length and breadth of the house together, and the product by the depth, which will give the cubic feet of the bulk of corn; then divide this last product by 12, and the quotient will be the number of barrels of shelled corn in the house or crib. If there be a remainder after the division, it will be so many twelfths of a barrel of shelled corn over.

Example.—12 feet long.
11 feet broad.
———
132
6 feet deep.
———
12) 792 cubic feet.
66 barrels of shelled corn.
5 bushels in a barrel.
———
330 bushels of shelled corn.

RULE FOR DETERMINING THE CONTENTS OF CISTERNS.—A simple rule to determine the contents of a cistern, circular in form, and of equal size at top and bottom, is the following: Find the depth and diameter in inches; square the diameter, and multiply the square by the decimal .0034, which will find the quantity in gallons for one inch in depth. Multiply this by the depth, and divide by 31½, and the result will be the number of barrels the cistern will hold.

For each foot in depth, the number of barrels answering to the different diameters are—

For 5 feet diameter, . . 4.66 barrels.
    6   "   . . 6.71 "
    7   "   . . 9.13 "
    8   "   . . 11.93 "
    9   "   . . 15.10 "
   10 "   . . 18.65 "

By the rule above given, the contents of barn yard cisterns and manure tanks may be easily calculated for any size whatever.

Long Measure—Is used to measure distances, and to ascertain the length of any thing, without regard to breadth.

10 lines, *l.*, make 1 inch, *in.*
12 inches   " 1 foot, *ft.*
3 feet   " 1 yard, *yd.*
$5\frac{1}{2}$ yards   " 1 rod or pole, *p.*
40 poles, or 220 yards, make 1 furlong, *fur.*
8 furlongs make 1 mile, *M.*
3 miles   " 1 league, *L.*
60 geographic, or $69\frac{1}{2}$ statute miles make 1 degree, *Deg.*
360 degrees make the circumference of the earth.

Twelve lines make an inch in France.

In measuring the height of horses, the *hand* (four inches), is used; and in measuring the depth of water, the *fathom* (6 feet), is used.

Surveyor's Measure.—This measure is used is ascertaining the length and breadth of land, roads, etc.

7 92-100 inches, *in.*, make 1 link, *l.*
25 links   " 1 pole, *p.*
4 poles, or 100 links, " 1 chain, *c.*
10 chains   " 1 furlong, *fur.*
8 furlongs   " 1 mile, *M.*

Mile.—The following exhibit of the number of yards contained in a mile in different countries, will often prove a matter of useful reference to readers:

FARMERS' DEPARTMENT.

A mile in England or America,   1760 yards.
Russia,   1100 "
Italy,   1476 '
Scotland and Ireland,   2200 "
Poland,   4400 "
Spain,   5028 "
Germany,   5066 "
Sweden and Denmark,   7223 "
Hungary,   8800 "
League in America or England,   5280 "

TABLE OF LEGAL WEIGHTS.—Showing the number of pounds which constitute a bushel in the following States:

ARTICLES.	Ill.	Iowa.	Wis.	Mich.	Ind.	Mo.	N.Y.	Ohio.
Barley	48	46	48	48	48	48	48	48
Bran	20	20	20	20	..	20	..	20
Broom Corn Seed	46	46	46	46	46	46	46	30
Buckwheat	52	52	40	42	50	52	48	52
Castor Beans	46	46	46	46	46	46	46	46
Charcoal	22	22	22	22	22	22	22	30
Coke	40	40	..	..	..	..	..	32
Corn (shelled)	56	56	56	56	56	56	56	56
" (in ear)	70	70	70	70	70	70	70	68 old. 70 new.
Corn meal	48	48	48	50	50	50	50	50
Dried apples	24	24	28	28	25	24	22	25
" peaches	33	33	28	28	33	33	32	33
" " pared	40	33	28	28	33	33	..	..
Flax Seed	56	56	56	56	56	56	56	56
Grass Seed, Blue	14	14	14	14	14	10	15	10
" " Clover	60	60	60	60	60	60	60	62
" " Hungarian	48	48	48	48	48	48	48	50
" " Millet	50	45	50	50	..	50	..	50
" " Orchard	14	14	..	14	..	14	14	14
" " Red Top	14	14	14	14	14	14	14	14
" " Timothy	45	45	46	45	..	45	45	45
Hemp Seed	44	44	44	44	44	44	44	42
Lime (unslaked)	80	80	80	80	80	80	80	80
Malt	38	36	38	38	38	38	34	34
Oats	32	33	32	32	32	35	32	33
Onions	57	57	57	57	57	57	57	56
" (top)	28	28	28	28	28	28	28	25
Osage Orange Seed	33	32	..	33	..	36	36	33

Table of Legal Weights and Measures—(Continued.)

ARTICLES.	Ill.	Iowa.	Wis.	Mich.	Ind.	Mo.	N.Y.	Ohio.
Peas	60	60	60	60	60	60	60	60
Plastering Hair	8	8	8	8	8	8	8	8
Potatoes, Irish	60	60	60	60	60	60	60	60
" Sweet	55	55	55	55	55	50	55	55
Rye	56	56	56	56	56	56	56	56
Salt (coarse)	50	50	50	50	50	50	50	50
" (fine)	55	50	56	56	50	50	56	50
Stone Coal	80	80	..	80	70	..	80	80
Turnips	55	55	55	55	55	55	55	60
Wheat	60	60	60	60	60	60	60	60
White Beans	60	60	60	60	60	60	60	60

DISINFECTANTS, AND HOW TO USE THEM.—The following is a copy of a card upon disinfectants just issued by the Board of Health of New York, together with directions for their use:

1. Quicklime, to absorb moisture and putrid fluids.—Use fresh stone lime, finely broken; sprinkle it on the place to be dried, and in damp rooms place a large number of plates filled with the lime powder. Whitewash with pure lime, and not with kalsomine.

2. Charcoal powder, to absorb putrid gases.—The coal must be dry and fresh, and should be combined with lime. This compound is the "clax powder."

3. Chloride of lime, to give off chlorine, to absorb putrid effluvia, and to stop putrefaction.—Use it as lime is used, and if in cellars or close rooms the chlorine gas is wanted, pour strong vinegar or diluted sulphuric acid upon your plates of chloride of lime occasionally, and add more of the chloride.

4. Sulphate of iron (copperas) to disinfect the discharge from cholera patients, to purify privies and drains.—Dissolve ten pounds of the copperas in a common pailful of water, and pour a quart or two of this strong solution into the privy, water-closet or drain, every hour, if cholera discharges have been thrown into those places; but for ordi-

nary use, to keep privies and water-closets from becoming offensive, pour a pint of this solution into every water-closet, pan, or privy seat, every night and morning. Always sprinkle a cupful of chloride of lime or lime powder in the same place and at the same time. Bed-pans and chamber-vessels are best disinfected in this way, by a spoonful of chloride of lime, and a spoonful of the copperas solution.

5. Per-manganate of potassa—to be used in disinfecting clothing and towels from cholera and fever patients, during the night, or when such articles cannot be instantly boiled. Throw the soiled articles immediately into a small tub of water, in which there has been dissolved an ounce of the Per-manganate salt to every six or eight gallons of water. A pint of Labarraque's solution of Chlorinated Soda may be used for the same purpose in the tub of water. Either of these solutions may be used in cleansing the oiled parts of the body of sick or dead persons. May also be used in bed-pans.

For water-closets use 4 and 3; privies, 4, 3, and 2; bed-pans and close stools, 1, 3, and 5; cellars, 1, 2, and 3; vaults and stables, 1 and 2, or 3 and 4, or any powders of coal tar.

For disinfecting soiled clothing, bedding and carpets, boil whatever can be boiled, if the articles have been soiled by cholera discharges. Use solutions of chloride of lime, or chlorinated soda, a quart of either solution to ten gallons of water, if the articles are coarse and their color of no consequence; but upon fine clothing that has been soiled in cholera or fevers, use the articles described under No. 5, in the list above.

In sick rooms use 1, 2, or 3; ventilate the bed-rooms, cleanse and dry the closets, ventilate the bed and bedding frequently in the sun.

Finally, let fresh air and sunlight purify every place and thing they can reach. Open and dry your cellars and vaults. Flush the water-closets and drains daily before throwing in the disinfectants as directed on this card. Let there be no neglect of domestic personal cleanliness.

CHOLERA—*Precautions Against it.*—The following is the address of the New York Metropolitan Board of Health:

The Board of Health publishes this simple statement, and beg the public to give it their earnest attention.

Cholera is generally a preventable disease, and in its early stages can be arrested, if the habits be good. Study, therefore, temperance in eating and drinking. Do not believe that alcoholic stimulants are useful in guarding you against an attack. Let the food be nutritious, and keep the digestive organs in a healthful condition. Use no stale or uncooked vegetables. Let your meat be fresh, and your vegetables well cooked, and all fruit fresh and ripe.

Cleanliness of the body is of the first consideration. Keep the skin in a healthy state by bathing the whole body, with a free use of soap. Cold bathing is best used in the morning—never just before going to bed. Dry frictions, or the warm bath may be more safely used just before going to bed.

Cleanliness in your homes is of equal importance. Let your apartments be dry—never damp. Suffer no decayed vegetables or stagnant water to remain in your cellars or yards. Any disagreeable smell from privies, cesspools, or sinks, is a proof of their unhealthfulness. Remove them by necessary repairs, lime, chloride of lime, or whitewashing. Ventilate well your houses and apartments. Expose your bedding to the air and sun. Avoid excessive fatigue. Keep regular hours in eating and sleeping. Wear flannel next to the skin. A good plan is, if the bowels are at all disordered, to wear a broad band of flannel (a flannel belly band) around the body, reaching from the hips to the ribs. Maintain the natural temperature of the body by sufficient clothing—especially keep the feet warm. Never, when heated, sit on the grass, or stone seats, or sleep under an open window; if exposed to wet, change your boots and clothes as soon as possible.

Take no purgative medicines, except by direction of a physician.

Cholera is almost invariably preceded by a diarrhœa, and this is in all cases to be promptly treated.

When diarrhœa is present, go to bed and maintain a position on the back, use abundance of blankets, and send for a physician.

A physician can always be obtained by applying at the nearest police station.

Stay in bed until you are well. Do not consider yourself well until you have had a natural movement from the bowels. Abstain from all drinks. Apply strong mustard plasters to the bowels.

In the absence of a physician, the adult can take ten drops of laudanum, and ten drops of spirits of camphor. A child of ten years may take five drops of laudanum and three of spirits of camphor. A child of five, and under, may take three drops of laudanum and three of spirits of camphor, and these drops may be repeated every twenty minutes, so long as diarrhœa, or pain, or vomiting continues. This will save time; but in all cases send for a physician.

Do not get up to pass the evacuations, but use the bedpan, or other conveniences.

Never chill the surface of the bed by getting out of bed.

Remove immediately all the evacuations from your rooms; scald all the utensils used, or disinfect them with chloride of lime; scald also your soiled clothing.

A SEA CAPTAIN'S REMEDY FOR CHOLERA.—Mr. G. S. Peabody, master of the packet ship Isaac Wright, has written a letter giving an account of the treatment of cholera cases which occurred on his vessel in January last, during a trip from Liverpool to New York. Captain Peabody says that within forty-eight hours after sailing, cholera appeared, and in ten days twenty-seven passengers had died of it, though they were treated "by the book." The Captain then applied a method of treatment that had been recommended by his predecessor in command, and did not lose another patient on that voyage, or since. The remedy was this: A teaspoonful of salt, and a tablespoonful of red pepper in a half pint of hot water. The captain says he was himself attacked by violent cholera, with cramps, etc., but the medicine "carried him through." He adds: "The medicine acts quickly as an emetic, say in one or two minutes. It brings up a very offensive matter, which sticks like glue. It was given, among others, to one old woman, eighty-four years of age, who was on deck—though weak, of course—the very next day. I have known it to be successfully used on board their ships by at least a dozen shipmasters besides myself. Its use is quite general in Liverpool, where even some of the regular doctors find it to their advantage to resort to it. Provided with this simple receipt, I no longer consider the cholera an unmanageable disease."

PETROLEUM FOR ASTHMA.—A correspondent of the "Country Gentleman" writes to that journal: "I have a son, six years old, that had the asthma in the most distressing form for three or four months, when he was one or two years old. We tried everything we could hear of without getting relief, till we were told to rub his neck and breast with petroleum, and we used it both crude and refined, experiencing very speedy relief, and a final and permanent cure, for he has not since had a return of it, and he is now a very healthy child."

LEMON JUICE TO RELIEVE PAIN.—Dr. Brandini, of Milan, says that lemon juice, or a solution of citric acid, relieves the pain of a cancer when applied to the sore as a lotion. The discovery was made accidentally, and the value of the application was confirmed by repeated experiments.

For Inflammation of the Kidneys.—Bathe the small of the back with sweet oil, and drink freely of balm tea.

Vomiting.—To stop vomiting, take gum camphor, pound it, pour on boiling water, and let the patient drink a tablespoonful every ten minutes, sweetened. Or, take a handful of green wheat or green grass, pound it, pour a little water on it, press out the juice, and let the patient drink a tablespoonful once in ten minutes. A tea of smartweed is also good.

On the True Signs of Death.—Dr. Descamps, of Milan, has presented to the French Academy of Medicine a memoir on the real signs of death. He draws the following conclusions, intended to guide public authorities in the precautions that should be taken against the danger of interring, prematurely, persons not really dead.

1. A greenish blue color, extending uniformly over the skin of the belly, is the real and certain sign of death.
2. The period at which this sign appears, varies much; but it takes place in about three days, under favorable circumstances of warmth and moisture.
3. Though discoloration of various kinds, and from various causes, may occur in other parts, the characteristic mark of death is to be found only in the belly.
4. Apparent death can no longer be confounded with real death, the belly never being colored green or blue in any case of the former.
5. This coloring of the belly, which may be artificially hastened, entirely prevents the danger of premature interment.
6. There is no danger to public health from the keeping of a body until the appearance of the characteristic sign of death.

www.ingramcontent.com/pod-product-compliance
Lightning Source LLC
Chambersburg PA
CBHW021803230426
43669CB00008B/622